Super Yachts

Cruising with Power and Style

The Deutsche Nationalbibliothek lists this publication in the Deutsche Nationalbibliografie; detailed bibliographical data are available on the internet at http://dnb.d-nb.de.

ISBN 978-3-03768-093-3

www.braun-publishing.ch

1st edition 2012

Selection of projects and layout: Sibylle Kramer
Editorial staff: Helen Gührer, Rossella Mungari Cotruzzolá, Meike Wittjohann
Text editing: Dagmar Glück
Translation: Geoffrey Steinherz
Art direction: Michaela Prinz, Berlin

Sibylle Kramer

Super Yachts

Cruising with Power and Style

BRAUN

Content

Super Yachts

Motor yachts embody luxury, comfort and boundless freedom. Their design and furnishings are the occasion for premium quality design. As a kind of floating mansion, yachts are governed by their own set of rules which pose a special challenge for designers. All the materials used must be able to withstand wind and weather and adapt to the physical conditions of the ocean.

Super Yachts presents a selection of 40 exclusive motor yachts which have been launched in the last few years. They run the gamut from relatively affordable series models which have been fashioned into individualized crafts by the creativity of designers, to the custom built, 150-meter-long, mega yachts of the luxury class. The market for these jewels is a small one. Internationally, there are only 5,400 yachts longer than 24 meters, which place them in the category of the super yachts.

The original nautical experience is to navigate the oceans solely with the force of the winds. However, most yacht owners prefer "cruising" or the breakneck speed of the power boat to sailing. Indeed the ratio of motor to sailing yachts has remained constant over the years at about 80:20. However, engineers are increasingly devoting their attention to ecological aspects of motorized yachting. For instance there are now many models on the market which use solar cells to generate power, making the vessels independent of generators. But the hunt for the next superlative is also part of the trend. And of course length is a benchmark. In their annual report, Camper & Nicholsons have determined that in the last two years alone 13 new vessels have joined the over 70-meter-long club of the great yachts. Other categories are speed, height, width and range, in which new records have been set. Thanks to modern damping systems yachts like the "AB 116" can attain a top speed of more than 50 knots, without compromising on comfort at such sportive exertion. The "Wallypower 118" even reaches a top speed of 65 knots, but its design looks like it hearkens from another world.

Inner values and unusual furnishing also can attract attention. Some boats queue up along with vessels of their class in the boat garage. For smaller harbors one always needs the appropriate coaster. Other yachts have as their own "tender" a helicopter on board, in order not to compromise the owners' brief respite on board with a complicated access to their yachts. In such vessels, it is almost understatement, if one limits oneself to exclusive designer furniture or controls the multimedia entertainment or lighting on board with an iPod.

Whether it be a superlative or the elegant comfort on an eco yacht – the vessels presented in this volume have one thing in common: life on the water as a unique experience. That requires adapting current design trends from the mainland to the special conditions of the sometimes rough seas. The clear, purist lines of contemporary loft architecture do not always lend themselves to safety on board, which requires more rounding off of edges. Often designers and shipbuilders work hand in hand and find surprising, innovative solutions: sophisticated spatial concepts in limited quarters, clever details and flattering lines, which accord the wind, weather and waves their due and pay homage to the sun. *Super Yachts* is an opportunity to explore the maritime dreams and visions of the 21st century.

exterior view

AB 116' at sea

AB 116'

Viareggio, Italy

Design: AB Yachts - Fipa Group
Shipyard: AB Yachts - Fipa Group
Length: 36.2 m
Beam: 7.5 m
Max Speed: 50 kn
Cruising Speed: 44 kn
Year of completion: 2011

The performance statistics of the AB 116' are unequalled in the market. At a top speed of 50 knots, the 36 meter long yacht flashes like a bullet over the water. A combination of water jet drives and conventional power plants are the key. For increased stability and reduced weight, the most modern materials from the aerospace industry were used for the hull and superstructure. Noise and vibration were reduced to a minimum thanks to the lean hydrodynamics, guaranteeing maximum cruising enjoyment even in rough seas. The interior is designed as comfortable extension of the sportive exterior. Large scale window fronts lend a feeling of expanse to the decks. Natural materials like wenge and holm oak, combined with fabric and leather in subdued color tones complete this unique human environment on the ocean.

sun deck with spa

from above to below: salon, cabin

from left to right: cabin, floor

main salon

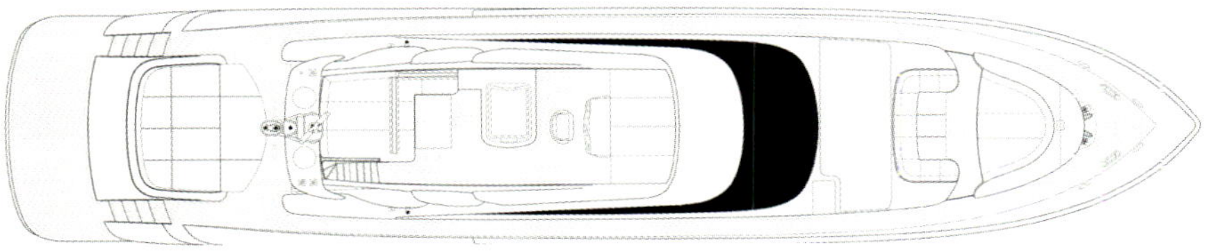

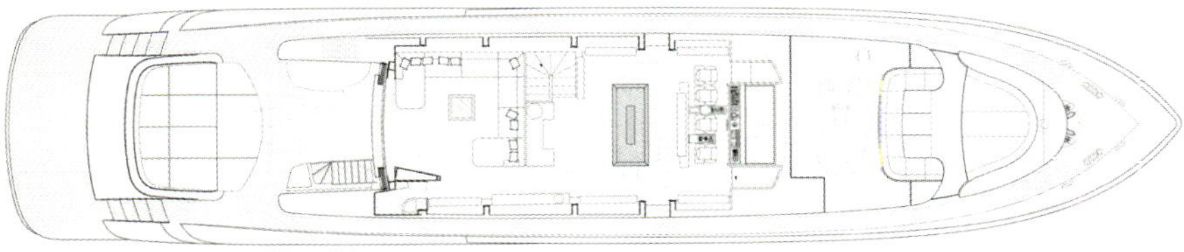

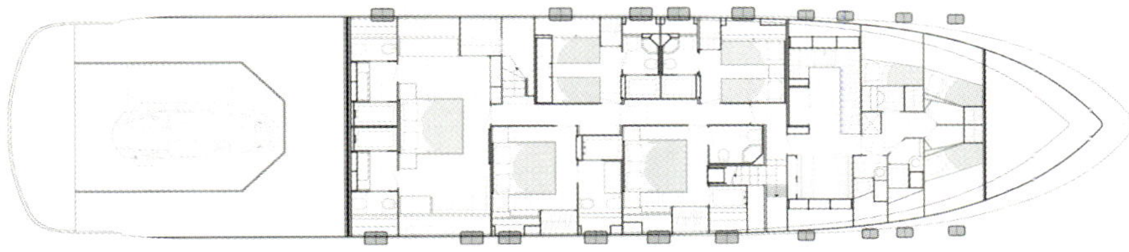

from above to below: flybridge, main deck, lower deck

from above to below, from left to right: main salon, dining area, wall detail

AB 140' at sea

AB 140'

Viareggio, Italy

Interior Design: Guido De Groot
Exterior Design: Giorgio Vafiadis
Shipyard: AB Yachts - Fipa Group
Length: 42 m
Beam: 8 m
Max Speed: 46 kn
Cruising Speed: 43 kn

The AB 140' is the product of a sophisticated hydrodynamic study with the goal of extracting unusual performance from a modest power plant. At the same time, extraordinary on-board comfort was to be maintained even under rough sea conditions. Thanks to the meticulous efforts of the builders, all expectations were surpassed: modern composites like Kevlar and carbon fiber reduce the total weight, enabling the AB 140' to reach a top speed of 46 knots. This yacht is from every standpoint a custom manufacture. The quality of the furnishings and an intelligent spatial design make it incomparable and one-of-a-kind. Black and purple set the tone in the interior, combined with panelling in wenge and olive wood. The result is an atmosphere of calm that can be enjoyed without distraction as the ocean whooshes silently by outside.

from above to below, from left to right: suite, sport area, sauna

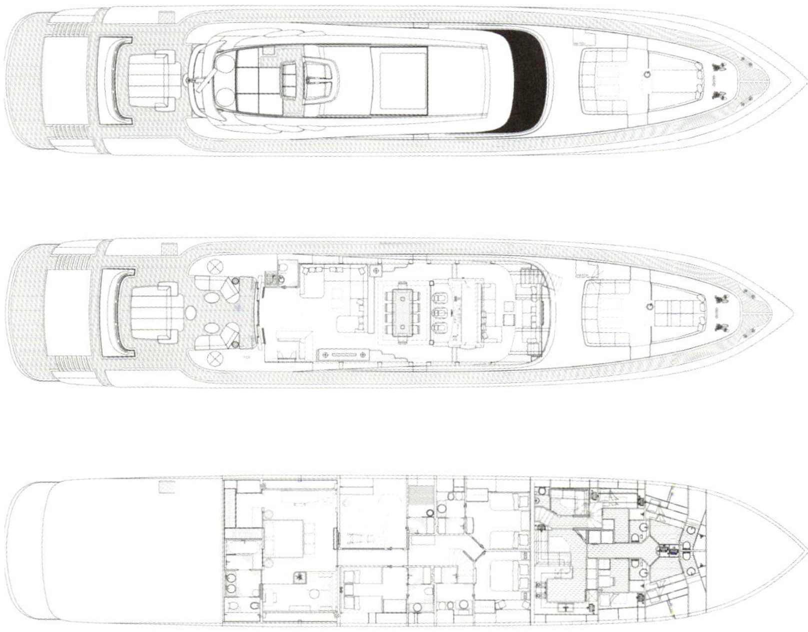

from above to below: flybridge, main deck, lower deck

guest cabin

stern view

from above to below, from left to right: exterior views, detail

Ac 74'

Enkhuizen, The Netherlands

Interior Design: Boris Levyant & Christiaan Bolinger
Exterior Design and Naval Architect: Olivier F. van Meer Design
Shipyard: Acico Yachts
Length: 22.9 m
Beam: 6.2 m
Max Speed: 24 kn
Cruising Speed: 12 kn
Year of completion: 2011

Fast, silent and something different – those were the parameters for the development of the AC 74' and that is exactly what emerges: whether it is pleasant cruising in calm seas or the power of moving quickly through high seas, its optimal maneuvrability and agility make every voyage with the AC 74' one of pure enjoyment. Even at a speed of 24 knots, the 26 meter long yacht glides almost soundlessly over the water. The interior allures with high quality furnishings and a clever spatial concept. In addition to the owner's suite, there are two guest cabins and two cabins for the crew. There is also room for a completely equipped galley . Warm materials like wood, leather and natural stone prevail in the interior. From the outside, the combination of supple elegance and classic motor boat design lend the AC 74' an out-of-the-ordinary appearance.

salon

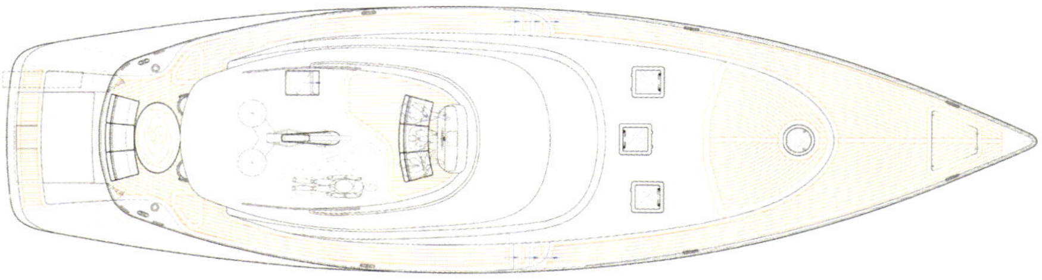

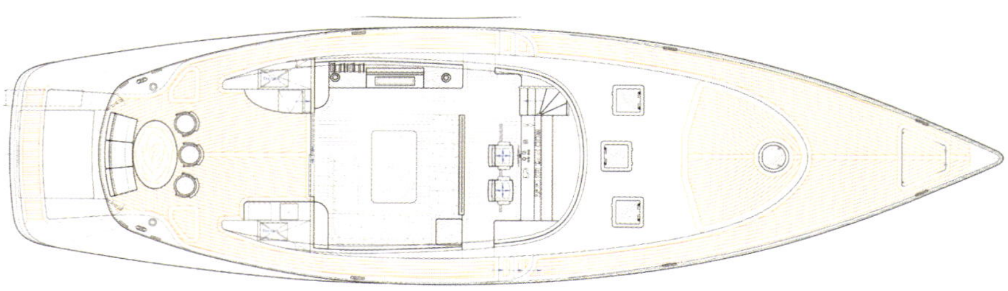

from above to below: upper deck, middle deck

bathroom

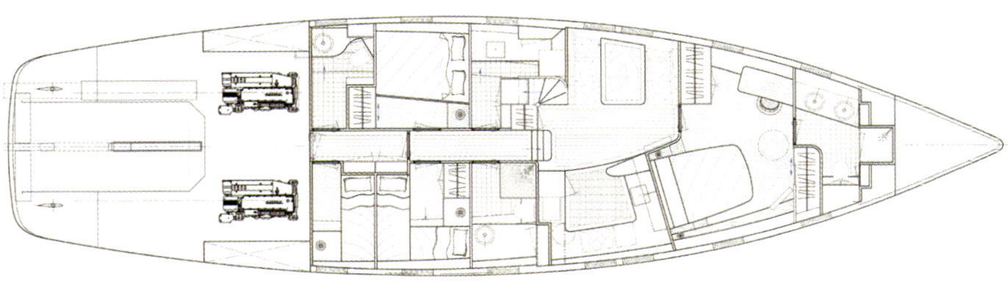

from above to below: profile, lower deck

stern view

Ancòra at sea

Ancòra

Cayman Islands

Design: Baglietto Design Department & Art-Line Interiors
Shipyard: Baglietto
Length: 42.9 m
Beam: 8.9 m
Max Speed: 16.5 kn
Cruising Speed: 15.5 kn
Year of completion: 2009

From the outside, the Ancòra is simply an elegant, sporty yacht. Inside, it is a jewel. The rooms were designed so that every area serves much more than a functional purpose. Each detail is part of a spectacular sculptural master plan. It includes the stairs which join the individual decks beneath a half transparent dome, the central lounge, which instead of fixed furnishings allows the passengers to arrange their own interiors according to mood, and the intimate fireplace contrasted with the expansive view of the panorama windows. A special highlight is the elegant relaxation couch made of soft leather. In the shape of a porthole, its stylish retreat is an invitation to gaze over the ocean, forgetting time and dreaming far beyond the horizon.

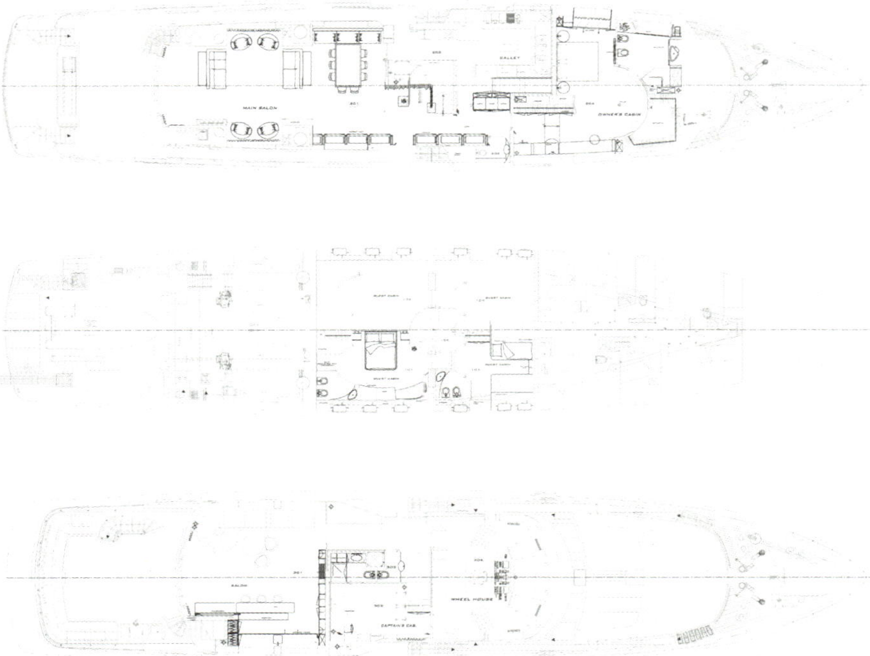

from above to below: main deck, lower deck, bridge deck

from left to right, fom above to below: bathroom, guest cabin, salon

from left to right, from above to below: salon, sun protection detail, salon

exterior view

salon

Arcadia 85'

Torre Annunziata, Italy

Design: Francesco Guida
Shipyard: Arcadia Yachts
Length: 25.9 m
Beam: 7.2 m
Max Speed: 19 kn
Cruising Speed: 16 kn
Year of completion: 2010

The Arcadia 85' silently follows the sun over the ocean. Literally, in a sense, because of the solar panels integrated in the double walled glass shell of the high-boarded yacht. At a rating of 4 KW, they take care of daily power needs on board, rendering generators superfluous. Not only do they obviate distracting noise and vibrations, the contribute to greater fuel economy and air quality.

The installation of a hybrid drive is also possible. The solar panels are an eye-catching element of the ceiling structure below deck and complement the clear design of the cabins in the Italian style by the Poltrona Frau Group. An environmentally compatible yacht with style, sailing beneath the sun of Arcadia.

Arcadia 85' at sea

from above to below, from left to right: sun terrace, dining area, kitchen

from left to right: guests cabin, owner's bathroom

owner's cabin

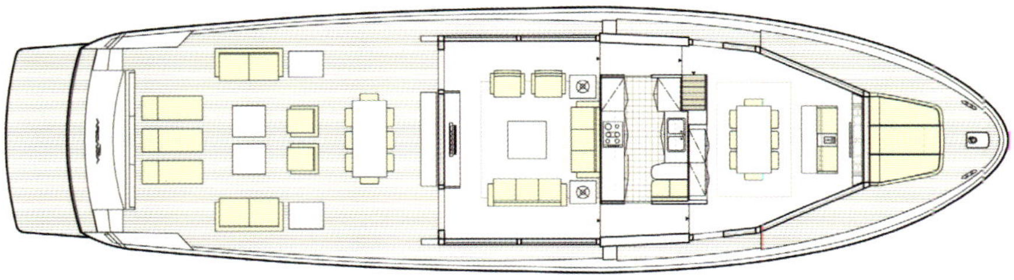

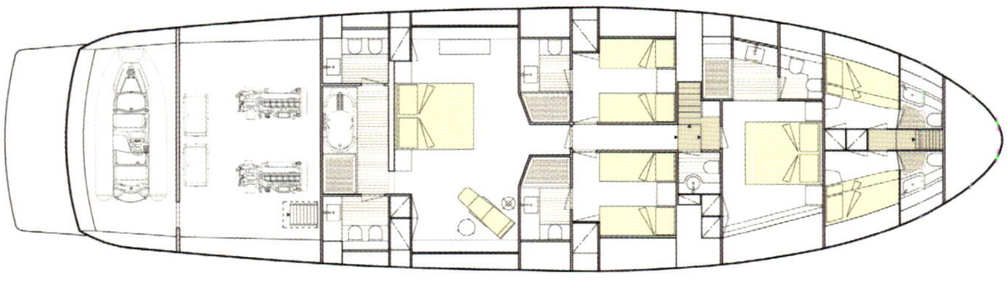

from left to right: main deck, deck with four cabins

from above to below: bow view, exterior view

stern view

Beluga

Antwerp, Belgium

Design: Puresang & Will Erens
Length: 100 m
Beam: 18 m
Max Speed: 26 kn
Cruising Speed: 20 kn
Year of completion: 2014

The Beluga design departs from convention in order to break new ground with an impressive mixture of whale boat, old fishing trawler and futuristic submarine. The two stories are marked by different variations of glass, binding them in a spectacular crystalline lustre. The forecastle encompasses a 650 square meter relaxation area with a pool and cocktail bar, with a greenhouse with palms and vegetable garden hidden away in the bow. The roof opens like the yawning mouth of a whale, revealing another 100 square meter of sundeck. In addition to the 20 cabins for the crew, there are five luxury suites and a suite for the owner, who may justly make his entrance like Captain Nemo in person.

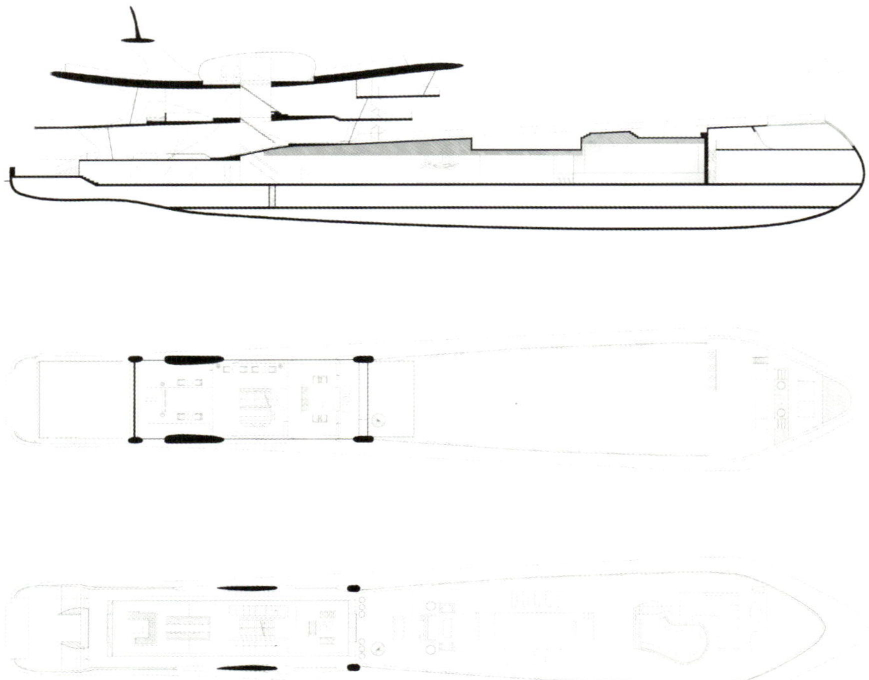

from above to below: section, upper deck, lower deck

salon

from above to below, from left to right: small living room, exhibition area, sleeping area

from above to below: spa area, exterior view

from above to right, from left to right: bow view, viewing platform, Big Fish at sea

Big Fish

Majuro, Marshall Island

Naval Architect: Gregory C. Marshall
Shipyard: Aquos Yachts
Length: 45 m
Beam: 9 m
Max Speed: 15 kn
Year of completion: 2010

Big Fish is angling to attract a new generation of admirers of custom made boats. Contemporary design combined with an unusually high standard of equipment and comfort make Big Fish the yacht for active individualists who relish the opportunity to remain at sea for several months at a time. With a range of over 9,000 sea miles, not a problem. A complete set of "aquatic toys" ensures that there will never be a dull moment: surfing equipment, diving gear, a pair of sailing dinghies and a universally deployable tender are on board. The multi-functional storage space in the bow can be alternately used as a boat garage, stowage for expedition equipment or a cabin. The decks offer flexible overnight possibilities and a spectacular view through the floor level windows. A big fish that is a real catch for the demanding adventurer.

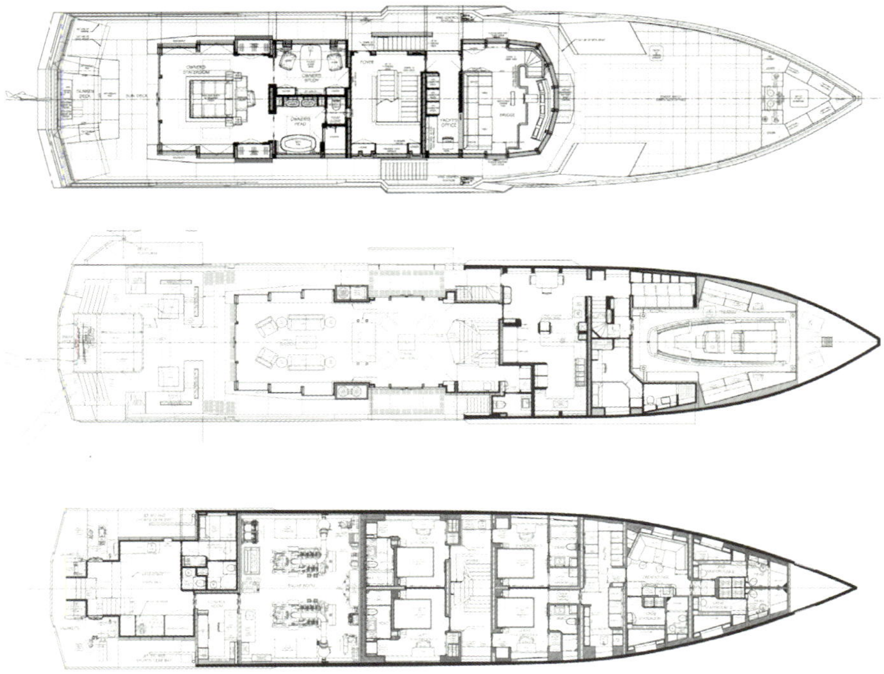

from above to below: upper deck, middle deck, lower deck

from above to below, from left to right: main salon, dining area, kitchen

from left to right, from above to below: guest cabin, bathroom, owner's cabin

from above to below, from left to right: main salon, dining area, stern view

Biscuit at sea

Biscuit

Hamilton, Cayman Islands

Interior Design: Hot Lab Yacht & Design
Exterior Design: CNC
Shipyard: Filippetti Yacht
Length: 28.5 m
Beam: 6.5 m
Max Speed: 25 kn
Cruising Speed: 22 kn
Year of completion: 2010

The 30 meter long Biscuit from the Filippetti dockyard is an exclusive design from bow to stern. Hot Lab Yacht & Design was responsible for the interior design and the furnishing of the sun deck. Two double cabins are located below deck: a VIP cabin and the owner's quarters. This area is marked by dark wood, high quality light leather coverings and the accent of steel. The comfortable salon has space for several sofas, a bar and the office. Light colored wall and ceiling colors underscore the great expanse of the space, highlighted by a painting by Paola Di Lusto. There is room for ten guests in the dining room, with a mess room and sleeping quarters below deck for a crew of four. The on-board kitchen features a wine cellar and is exclusively equipped with Gaggenau kitchen components. The multifunctional furniture for the sundeck designed by Hot Lab leaves no wish unfulfilled.

from above to below, from left to right: main salon, guest cabin, interior stairs

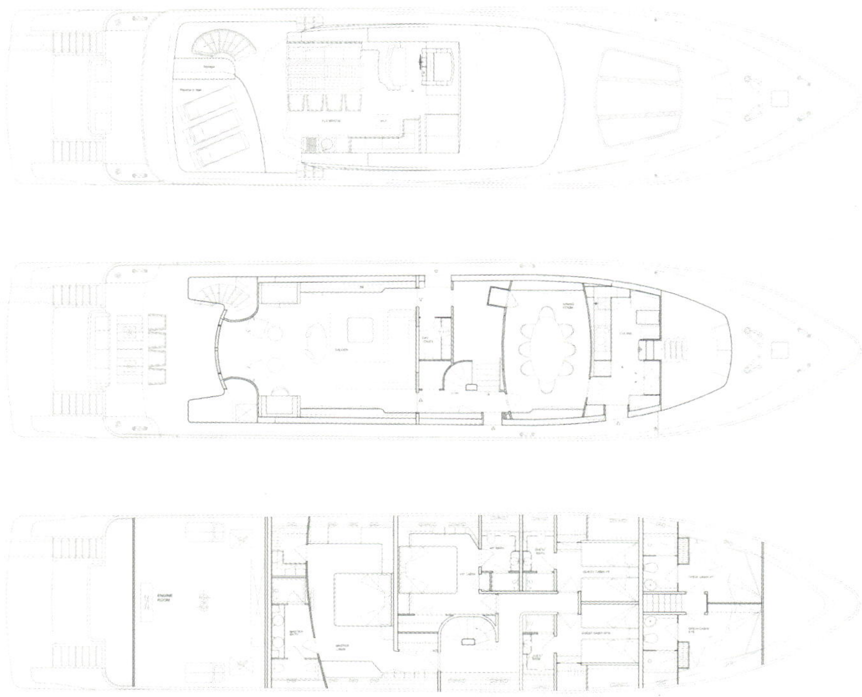

from above to below: flybridge, main deck, lower deck

from left to right, from above to below: workspace, VIP-cabin, master cabin

bow view

sun deck

Exuma

Viareggio, Italy

Design: Philippe Briand & Vitruvius Yachts
Engineer: Perini Navi
Shipyard: Perini Navi & Picchiotti
Length: 49.5 m
Max Speed: 16.5 kn
Cruising Speed: 12 kn
Year of completion: 2010

More than 500 years ago, Leonardo da Vinci created the "Vitruvian Man". The inspiration for this famous drawing was the theory of the "well proportioned man" formulated by the Roman architect Marcus Vitruvius Pollio fifteen hundred years before. Vitruvius is the inspiration for a new series of well proportioned yachts, whose proportions also play a significant role. The 50 meter long Exuma is the first in the Vitruvius line, which will comprise high performance motorized yachts from 44 to 73 meters in length. With low fuel consumption and a range of 4,500 sea miles, the Exuma is a true explorer. The aluminium hull is one of its secrets: at 330 tons, it has half the weight of similar yachts. The low center of gravity and the hydrodynamic contour of the shell set new standards for motorized yacht performance.

from left to right, from above to below: salon, guest cabin, owner's bathroom

Exuma at sea

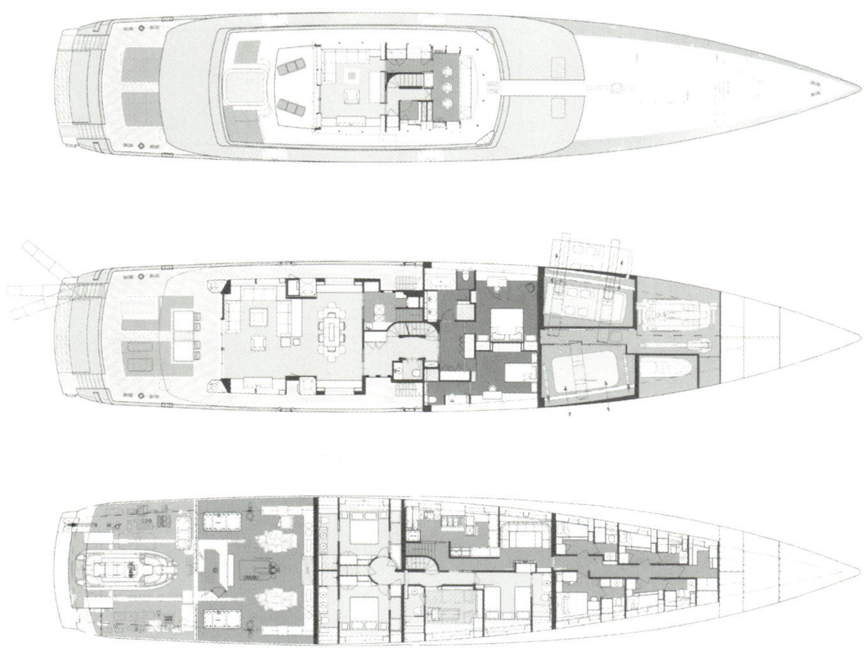

from above to below: bridge deck, main deck, lower deck

from above to below, from left to right: bow seating area, exterior view

Fjord 36' Open at sea

Fjord 36' Open

Greifswald, Germany

Design: Patrick Banfield / Allseas Design
Shipyard: Hanse Yachts
Length: 10.8 m
Beam: 3.7 m
Max Speed: 43 kn
Year of completion: 2011

Designer Patrick Banfield designs tenders for 130 meter long mega yachts and also designed the legendary Wally tender. The German Hanse Yachts AG was able to retain him for the Fjord 36' Open. At a first glance, one does not notice that the only eleven meter long boat is habitable. But a double bed and a comfortable shower bath have been installed beneath the control panel. The dining area is located on the foredeck. The Fjord glows with understatement and provides an exclusive alternative to all the boats developed for the mass markets with two cabins, bath, kitchen and a living room. With 500 horsepower, the engines accelerate the craft to an impressive 42 knots. The range can be extended to 500 sea miles with an auxiliary tank. The Fjord is a comfortable, practical and functional vessel without competition, and is also extraordinarily maneuvrable.

from above to below, from left to right: living space, bathroom, cabin

from left to right, from above to below: stern details, outdoor area

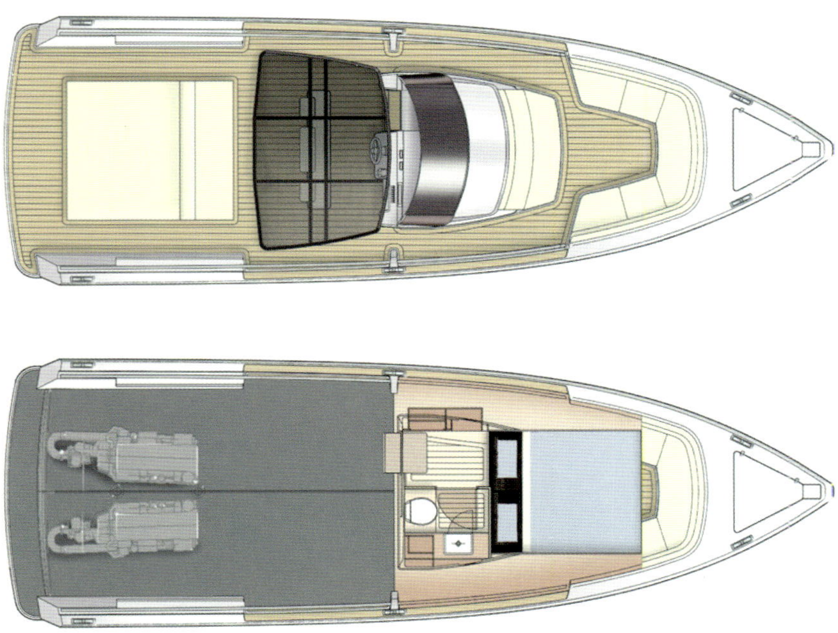

from above to below: upper deck, lower deck

Hedonist at sea

exterior view

Hedonist

Tivat, Montenegro

Design: Art of Kinetik
Shipyard: Art of Kinetik
Length: 19 m
Beam: 5.3 m
Max Speed: 40 kn
Cruising Speed: 33 kn
Year of completion: 2010

Passengers on the Hedonist know they can enjoy life on the water to the hilt. The open, 19 meter long yacht is offered in two luxurious models: for day outings and longer tours. The yacht is distinguished from the usual look by style, quality and comfort. High quality materials, open floor plans and modern living design create a tailor-made environment for the owner. The hull is made of wood, and is thereby something finer, timeless, and beautiful. Wood also has many functional advantages. It is more stable than most plastics and is not subject to fatigue like metal. It also absorbs noise and vibration. Like all yachts built by Art of Kinetik, the Hedonist will be a long-lived, floating masterpiece for generations to come.

from above to below, from left to right: salon, cabin, bathroom details

kitchen

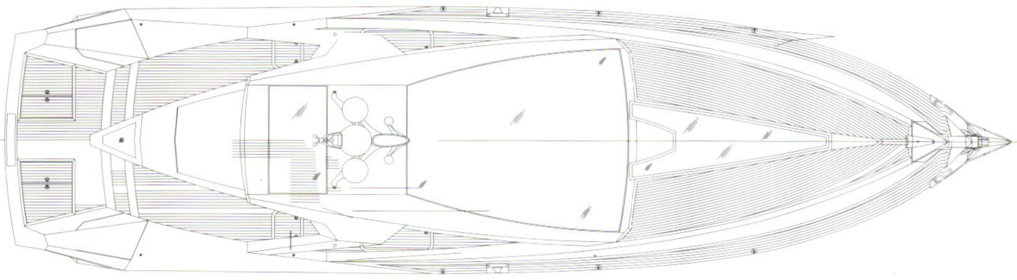

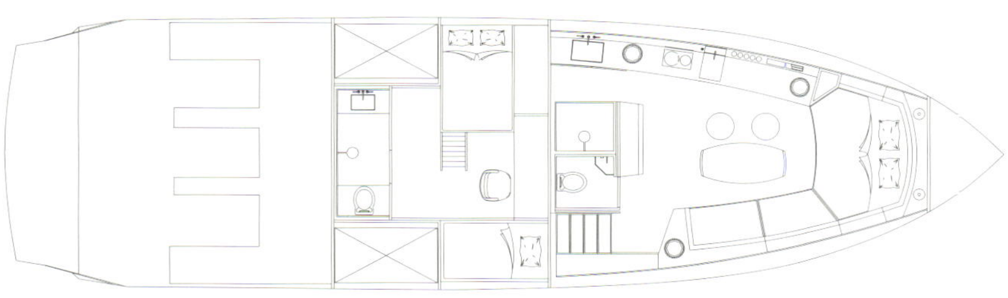

from above to below: upper deck, lower deck

exterior view

from above to below, from left to right: exterior view, sun deck, detail

H2OME

Valletta, Malta

Interior Design: A-Lab Marine
Exterior Design: Navirex
Length: 44 m
Beam: 8 m
Max Speed: 36 kn
Cruising Speed: 18 kn
Year of completion: 2010

The H2OME glides as swiftly as an arrow. At speeds of more than 40 knots, her elongated hull rises from the waves, her striking silhouette melting into the ocean. This organic connection of technology and the primal element is that which drives the designer to create modern living space in the midst of the maritime environment. The 360° panorama view in the completely glass-enclosed salon opens the upper deck to the ocean. In the interior, the interplay of nature and high-tech continues: the greened walls here provide an emotional contrast to the clear technical surfaces. The indirect lighting concept makes perfect the feeling of life in the weightlessness of water.

reading area

from above to below, from left to right: salon, dining area, detail

from above to below, from left to right: cabin, small salon, nautilus detail

from left to right: owner's bathroom, stairs

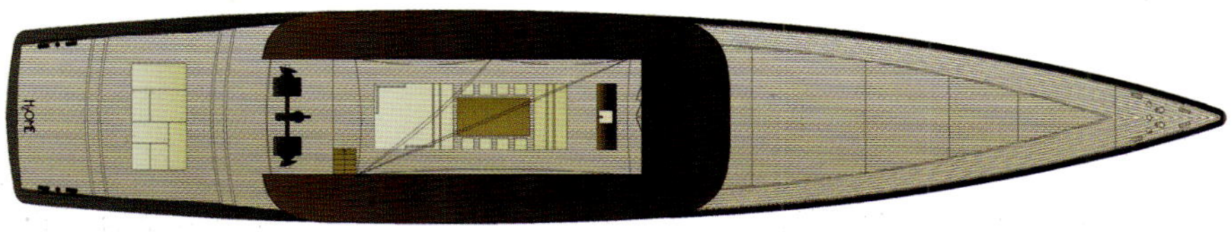

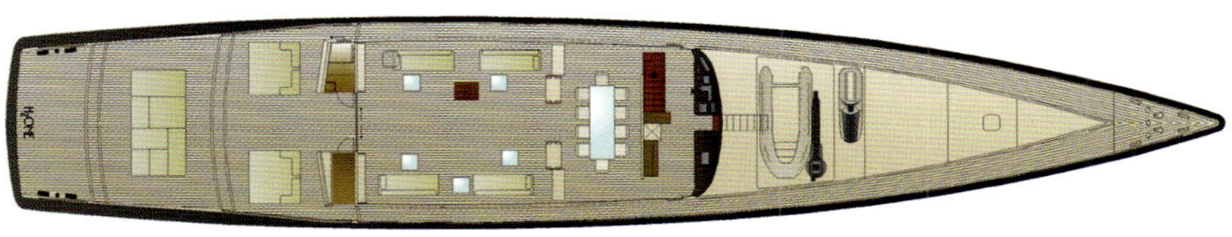

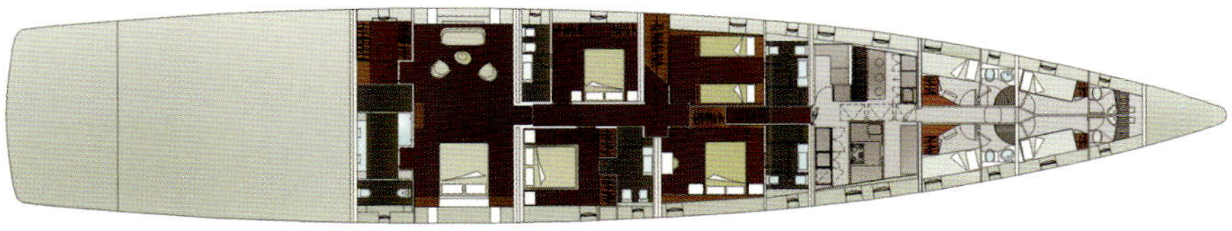

from above to below: sun deck, main deck, lower deck

from left to right: exterior view, stern view

Icon at sea

Icon

Harlingen, The Netherlands

Exterior Design: Redman Whiteley Dixon
Interior Design: Studio Linse
Shipyard: Icon Yachts
Length: 62.5 m
Beam: 11.40 m
Max Speed: 16 kn
Cruising Speed: 12 kn
Year of completion: 2009

The new Dutch shipyard, Icon Yachts, has quickly made a name for itself which stands for highest quality. The proof is the 62.5 meter Icon, which was created in close collaboration with Redman Whiteley Dixon. The designers are responsible for the external appearance, while the interior was conceived by Sudio Linse. In addition to the six large suites with space for up to twelve guests, the Icon can take on a 19 man crew. From the complete assortment of water sport equipment to the sophisticated spatial concept, this super yacht is in a position to spoil its passengers with every conceivable comfort. The guest cabins have direct access to the beach club, which, with Jacuzzi, Sauna and bar, leave no wish unfilled. A private stairway leads from the owner's suite to a viewing platform, appropriate for viewing romantics sunsets and enjoying nights beneath a starry sky.

from above to below, from left to right: exterior salon, bar, bow details

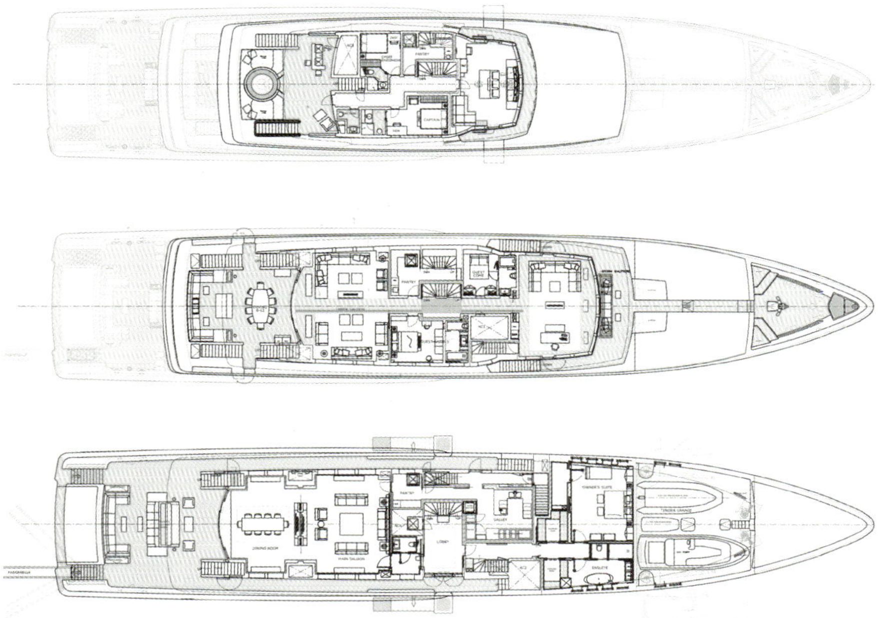

from above to below: bridge deck, upper deck, main deck

salon

from above to below, from left to right: salon, dining area, master cabin bathroom

from left to right, from above to below: guest cabin workspace, guest cabin bathroom, guest cabin

from above to below, from left to right: salon

from above to below: exterior view, owner's lounge

Import III

Amsterdam, The Netherlands

Interior Design: Egg and Dart Design Corporation
Exterior Design: De Birs Yachts / Barilani
Shipyard: De Birs Yachts
Length: 26 m
Beam: 6 m
Max Speed: 28 kn
Cruising Speed: 23 kn
Year of cpmletion: 2007

The serial production, 26 meter long Import III has been optimized in terms of form, function and design. Egg and Dart have innovatively reinterpreted the traditional inner life of the mega yacht. The main feature was the design of a large open salon. Cream colored leather, high quality textiles, glass partitions and metal parts made of rust free steel create an elegant and comfortable atmosphere. Each cabin is clearly distinguished by the use of the high quality materials. The characteristically shaped windows of the shell have been mirrored inwards in the lounge, allowing for a stupendous panoramic view. The efforts of designers have transformed a standardized yacht into a distinctive, one-of-a-kind vessel.

from above to below, from left to right: owner's cabin, workspace, bathroom

from left to right, from above to below: owner's bathroom, guest cabin, workspace

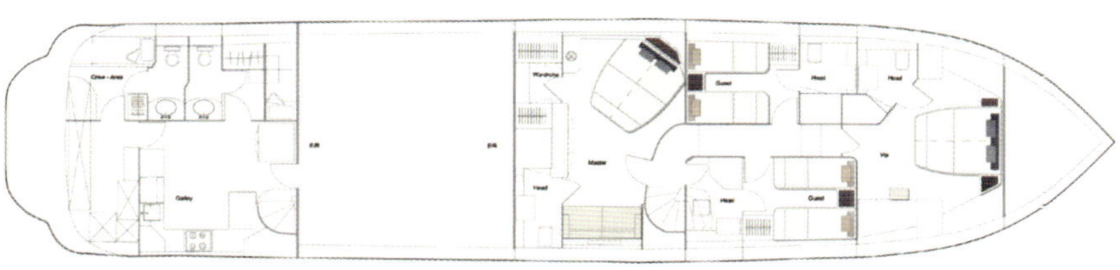

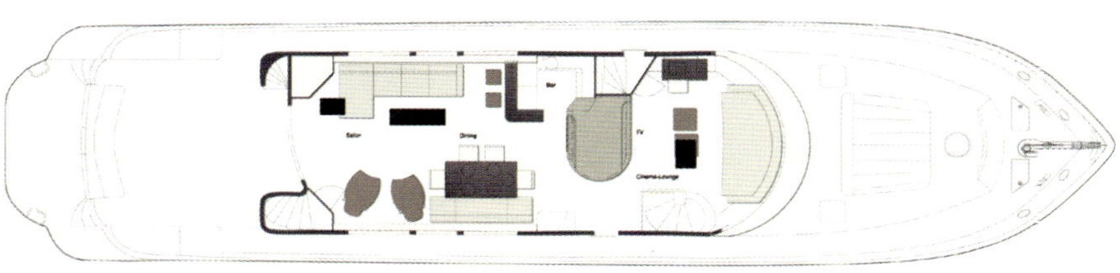

from above to below: lower deck, main deck

from above to below, from left to right: flybridge, sun deck

exterior view

Lifestyle

Madeira, Portugal

Interior Design: Egg and Dart Design Corporation
Exterior Design: Frank Mulder
Shipyard: DMC (Danube Marine Consultants)
Length: 36.6 m
Beam: 7.8 m
Max Speed: 27 kn
Cruising Speed: 15 kn
Year of completion: 2005

Elegance and sport distinguish the 36 meter long Lifestyle. A subdued language of form and the modern color and material concept combines exterior and interior in a coherent overall appearance. The clear, dynamic lines of the hull continue in the interior. Light, high quality materials harmonize with the blue tones of sky and ocean, resulting in an atmosphere that is both airy and luxurious. The sophisticated lighting concept is perfectly coordinated with the calming marine environment. Each cabin below deck is assigned its own theme, with colors ranging from beige to brown, to copper tones. Materials like leather, silk and wood round off the interior design, making the yacht into a swimming loft for all who can appreciate discretion in luxury.

dining area

salon

from left to right, from above to below: owner's cabin, owner's bathroom, guest cabin

from left to right: handrail detail, stairs to main deck

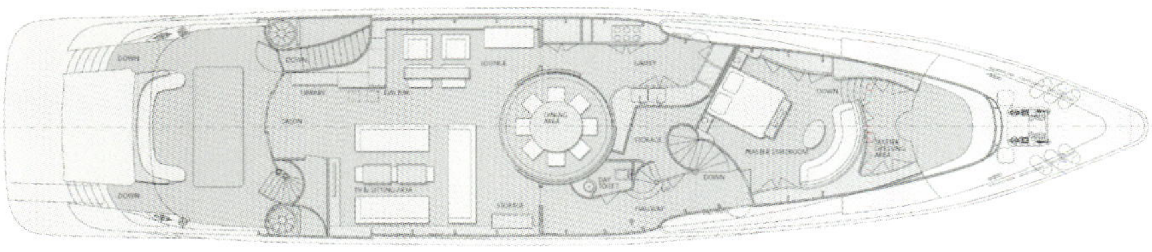

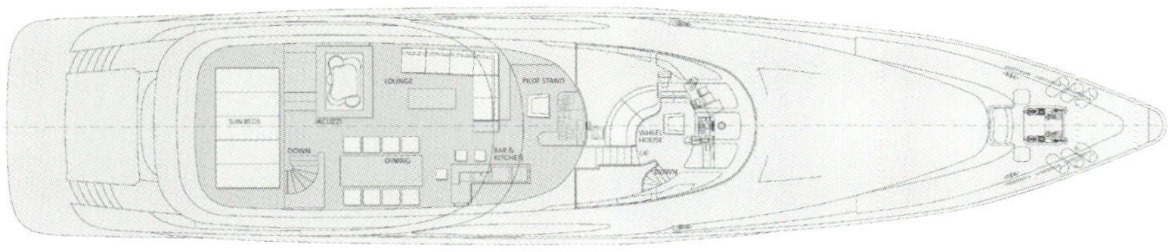

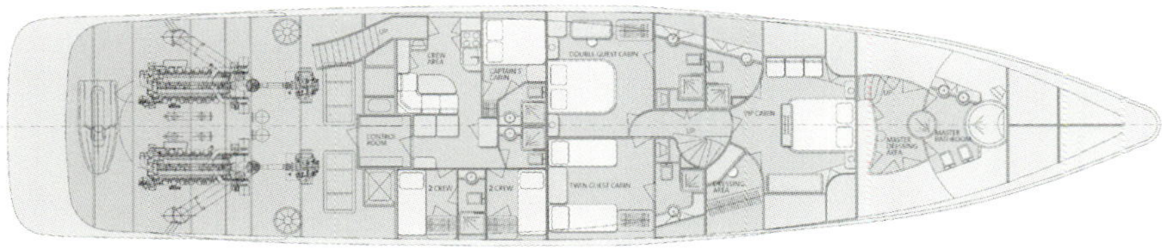

from above to below: main deck, upper deck, lower deck

from above to below, from left to right: dining area, upper deck, bar

exterior views

LR Noor

Valletta, Malta

Interior Design: Hot Lab Yacht & Design
Exterior Design: Bilgin Yachts
Shipyard: Bilgin Yachts
Length: 37.5 m
Beam: 7.5 m
Max Speed: 24 kn
Cruising Speed: 20 kn
Year of completion: 2010

From the outside, the 37 meter long LR Noor has lean lines – while the interior unfolds into a floating palace. Thassos marble, choice carpets and furniture contribute to the luxurious atmosphere. The flooring and panelling exude the warm wood tones of polished oak throughout the entire ship. Large windows let in natural light into all decks, discretely aided at certain strategic locations with spotlights and energy saving LEDs. While the cabins are distinguished by clear lines and a lighter color palette, the great salon makes its entrance in the rich tones of inlay work, cloth and upholstered leather. The salon floor is done in dark ebony. Another highlight is the dining area, where up to eight people may dine beneath a spectacular skylight.

main salon

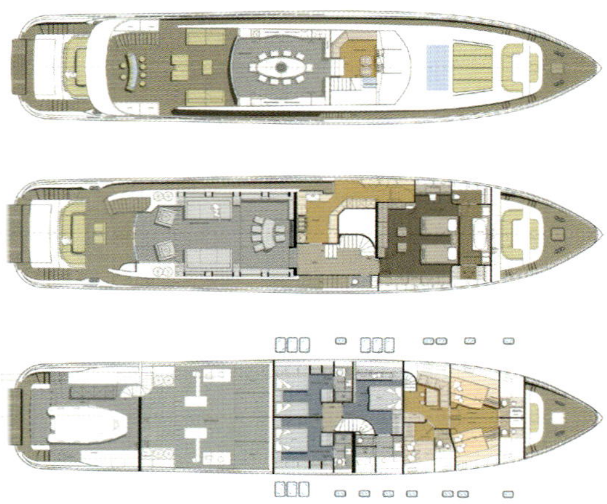

from above to below: upper deck, main deck, lower deck

from left to right, from above to below: owner's bathroom,
owner's cabin

from above to below, from left to right: cockpit with sun sail, cockpit without sun sail, exterior view

Magnum 60 at sea

Magnum 60' Tiranno

Porto Ercole, Italy

Design: Lazzarini Pickering Architetti
Shipyard: Magnum Marine
Length 18.3 m
Beam: 5.2 m
Max Speed: 50 kn
Cruising speed: 47 kn
Year of completion: 2006

In a unique way, at only 13 meters the Magnum 60' Tiranno combines comfort with sport. The designer masterfully fulfilled the customer's desire to install complete living quarters below deck in a sport yacht. While the external lines emphasize the dynamic, aggressive character of the yacht, especially ingenious solutions were required for the interior. For instance, the chaise lounges on the multi-functional sun deck can be reinvented as dinettes. Space below deck is also flexibly used. Sliding doors are enlisted as tables when the two separate sleeping cabins merge to form a complete living room with comfortable sofas. The tone-setting white in the interior suggests a feeling of expanse, with associations of a futuristic space capsule.

from above to below, from left to right: cabin with partitions opened, partitions turned into tables

from left to right, from above to below: cabin with beds and partitions closed, side panels closed, cabin with partitions fully opened

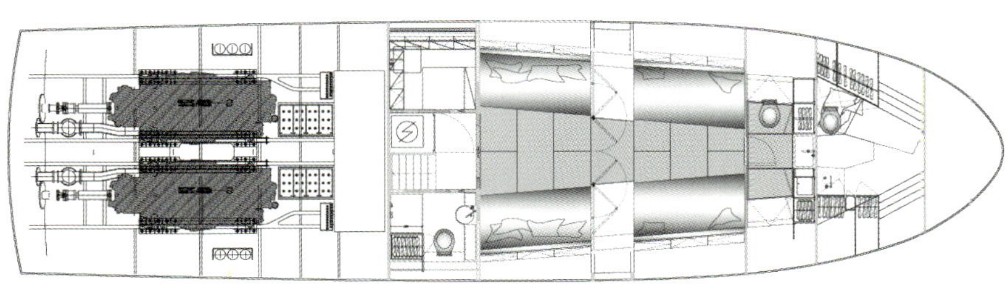

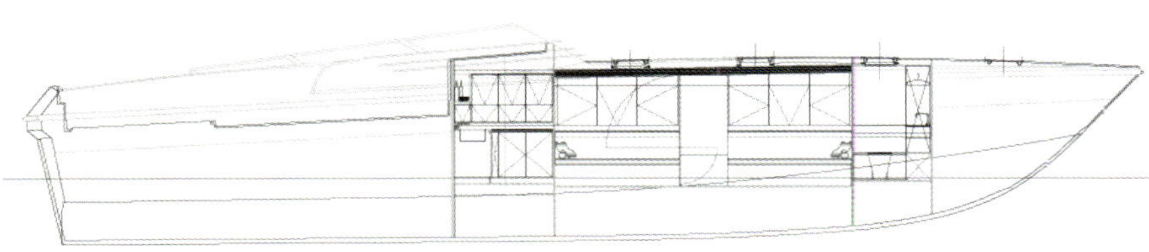

from above to below: deck, section

from above to below: main salon, exterior view

Manhattan 63' at sea

Manhattan 63'

Port Adriano, Mallorca

Design: Sunseeker Design
Shipyard: Sunseeker International Limited
Length: 21.1 m
Beam: 5.1 m
Max Speed: 33 kn
Cruising Speed: 23 kn
Year of completion: 2010

The Manhattan 63', the latest masterpiece of Sunseeker Shipyards, merges perfect modern design and innovative technology. The yachts may be equipped with a conventional power plant or with the Volvo Penta 1200 horsepower, which, with better performance, yields reduced fuel consumption and can be controlled with a joystick. That increases the maneuverability of the yacht considerably. At a length of 21 meters and range of 270 sea miles, the stylish Manhattan is an ideal family yacht, for comfortably cruising to the mooring of the day or ranging the open seas at a brisk top speed of 33 knots.

owner's suite

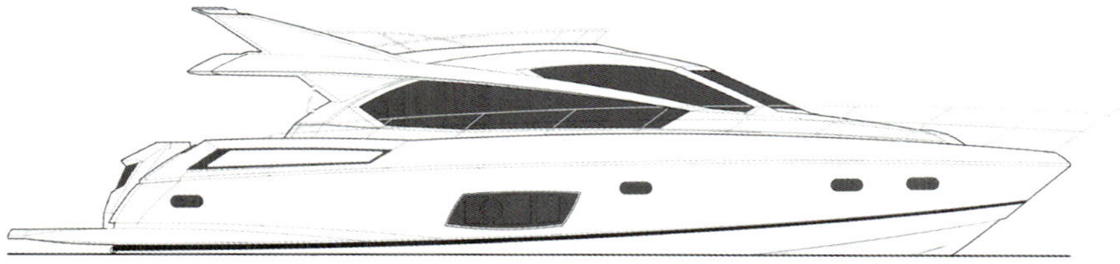

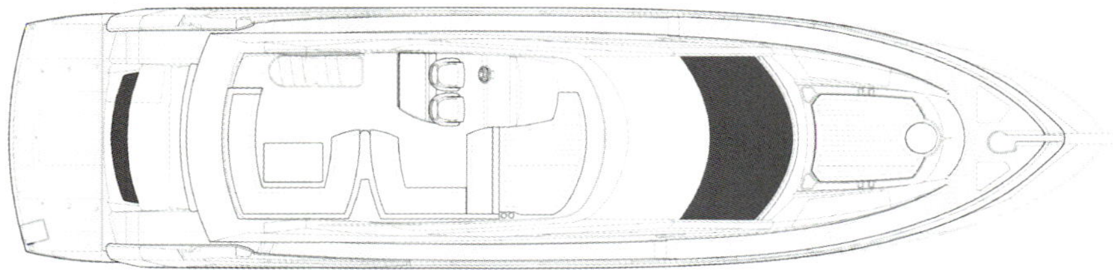

from above to below: profile, flybridge

from left to right, from above to below: bathroom, guest cabin, kitchen

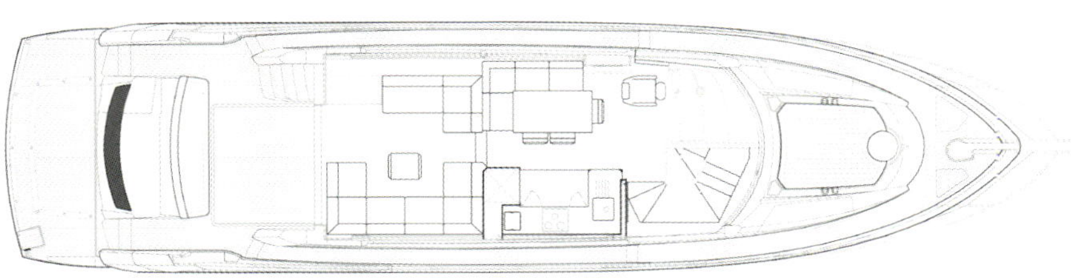

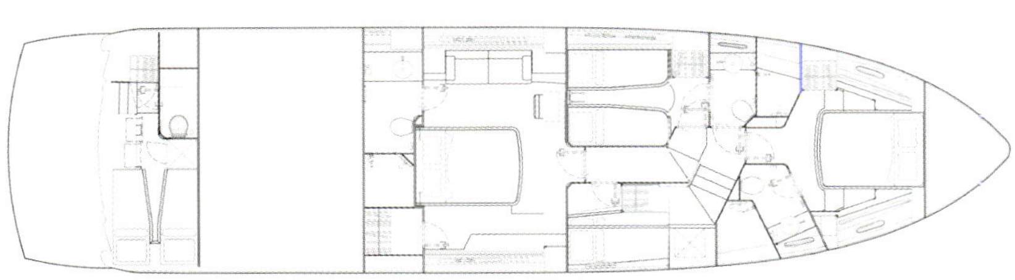

from above to below: main deck, lower deck

from above to below, from left to right: kitchen, dining area

exterior view

Mathisse

Fano, Italy

Interior Design: A-Lab Marine
Exterior Design: Stirling Design
Shipyard: Seanest Fano
Length: 25.6 m
Beam: 6.1 m
Max Speed: 30 kn
Year of completion: 2009

In close consultation, the Mathisse designers, architects and engineers developed an innovative concept. The result is a yacht with an aggressive outward appearance and an interior design which transmits a feeling of weightlessness. State of the art technology and modern materials are the key. The galley in the great salon disappears behind glass walls, which with the aid of liquid crystal technology can be made invisible. Four cabins and baths, the independent crew area and the boat garage fill out the spatial program. On the upper deck, along with an exercise area which includes a treadmill, there is a large sun deck and a dining space for up to eight passengers, who will experience unforgettable moments on the ocean.

from left to right, from above to below: workspace, bathroom, cabin

cabin

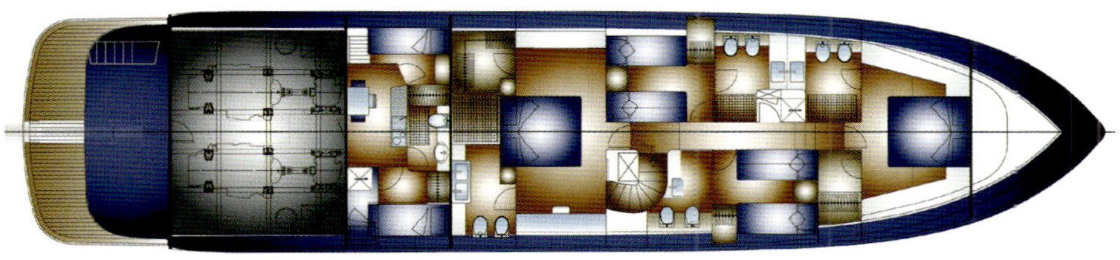

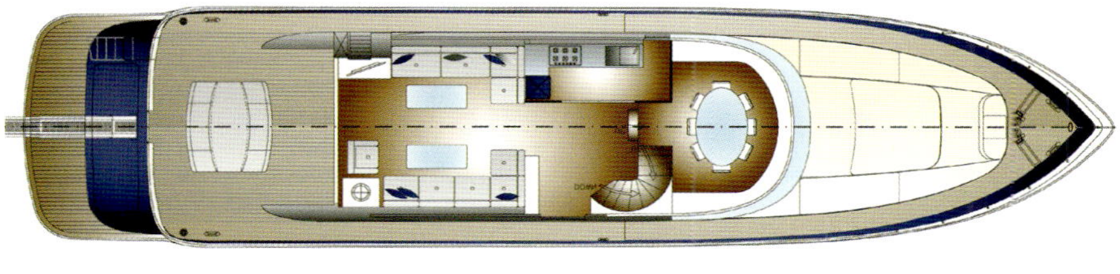

from above to below: main deck, sun deck

from above to below, from left to right: flybridge, salon, bar

Mystic at sea

Mystic

Antalya, Turkey

Interior Design: Schnaase Interior Design
Exterior Design: Diana Yacht Design
Shipyard: CMB
Length: 45 m
Beam: 9 m
Max Speed: 17 kn
Cruising Speed: 13 kn
Year of completion: 2010

The Mystic is the impressive result of the collaboration between the Turkish shipyard CMB and the German interior designer Schnaase. The basic idea of a mystical subject provides the yacht with its signature style, recognizable in the layout as well as the choice of materials. On the main deck, unusual for this location, is a spacious gym with high windows, modern fitness machines and a sauna. The owner's suite is then located on the upper deck – with a private terrace in front of the sleeping quarters. This concept consciously departs from the usual loft styles, producing a little oriental jewel. Expressive materials like Macassar wood and mother-of-pearl, and silk rugs beneath crystal chandeliers with accents of gold merge into a nouveau Art Deco, making the yacht into a kind of floating dream from the 1,001 Nights.

owner's cabin

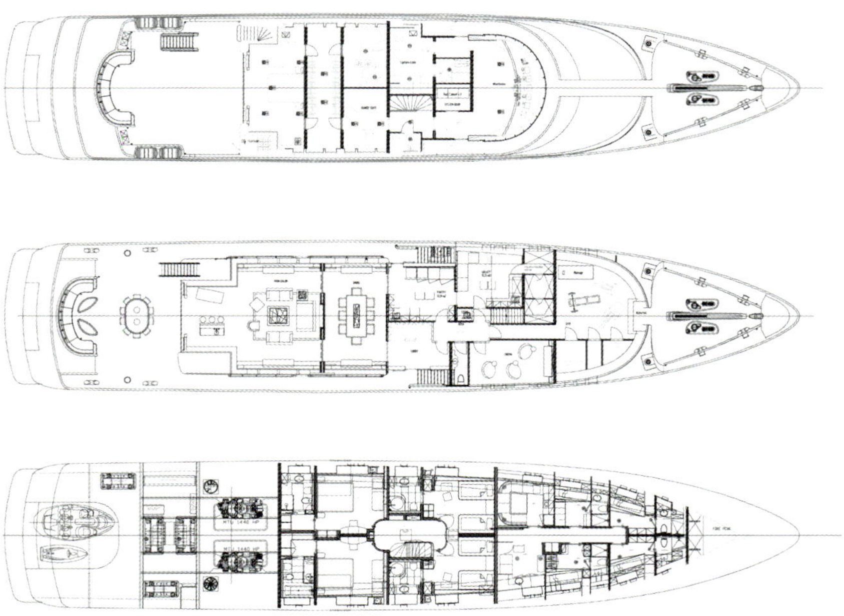

from above to below: sun deck, main deck, lower deck

from left to right, from above to below: guest cabin, owner's bathroom detail, owner's bathroom

bow view

stern view

Ocean Emerald

Isle of Man, United Kingdom

Design: Foster + Partners
Shipyard: Rodriquez Cantieri Navali
Length: 41 m
Beam: 8.4 m
Max Speed: 17.5 kn
Cruising Speed: 16 kn
Year of completion: 2009

The boats of the YachtPlus fleet redefine the exterior and interior of traditional yachts, thereby creating unique spaces which are as technologically advanced as they are elegant. The contemporary design allows for more than 30 percent more deck surface than that available on comparable yachts. This gain translates into luxurious cabins for up to twelve guests. Floor to ceiling windows located in the salon on the upper deck provide an unimpeded view all the way to the horizon. Natural materials like the teak furniture and the coconut fiber of the flooring blur the boundary between inside and outside. A glass enclosed stairwell allows natural light into the lower deck. Indirect lighting allows for a different mood in every room. Depending on the time of day, the windows look opaque or transparent.

dining area with large sliding doors to deck

cabin

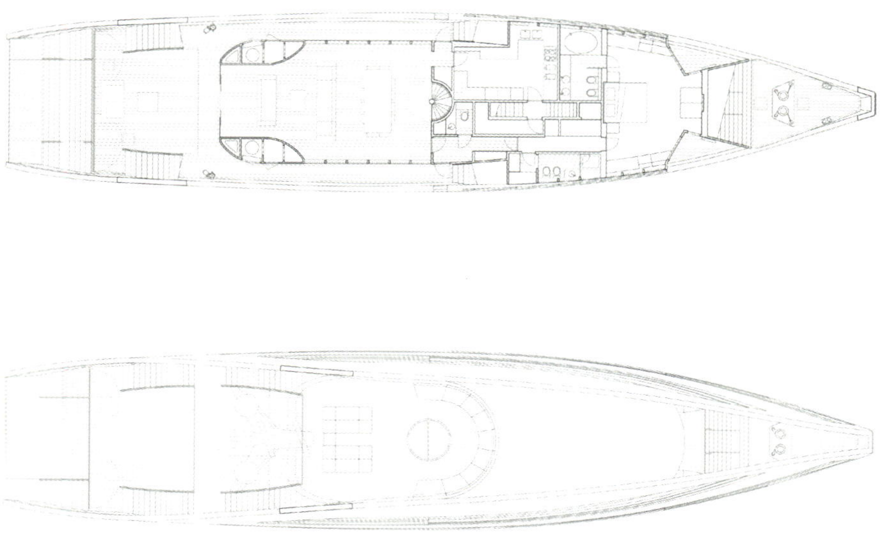

from above to below: main deck, sun deck

port s de view

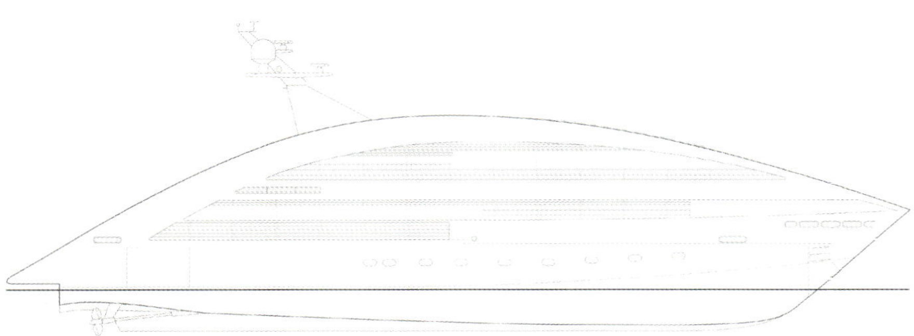

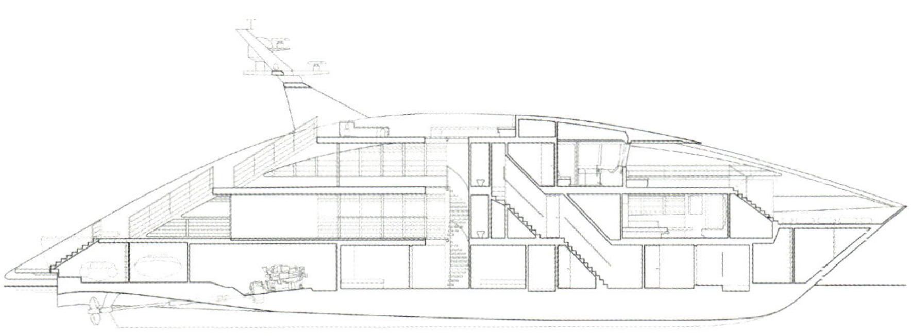

from above to below: profile, section

exterior view

Oculus at sea

Oculus

Design: Schöpfer Yachts
Length: 76.2 m
Beam: 12.8 m
Max Speed: 21 kn
Cruising Speed: 17 kn

The Oculus has both a sleek as well as futuristic aura. This unusual long-range yacht is the first design launch by Schöpfer Yachts and is designed to accommodate up to 12 passengers in style and comfort. The boat's shape, with its dramatically receding bow, recalls the skull structure of large fish and sea mammals. The low rider profile creates additional relaxation surfaces on deck which in high seas can be closed via retractable side panels. The futuristic design and generous dimensions of the interior spaces contribute to a feeling of weightlessness. This vessel houses a salon with a ceiling height of four meters, and the cylindrical dining rooms extends over two stories. The bathrooms and stowage are located in four freestanding tubes.

salon

salon

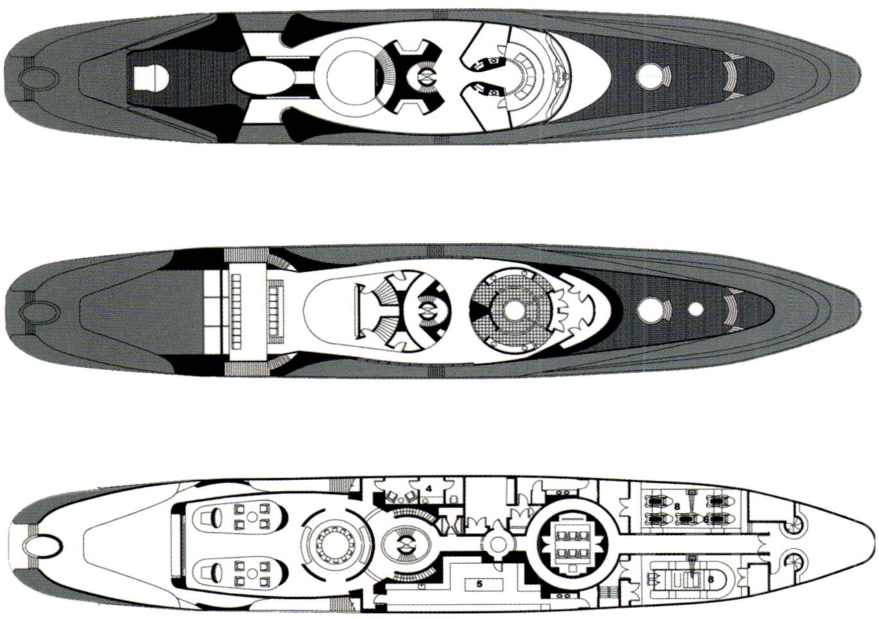

from above to below: upper deck, middle deck, main deck

exterior view

Pallacium at sea

Palladium

Cayman Islands

Design: Michael Leach Design Limited
Shipyard: Blohm & Voss
Length: 95.1 m
Beam: 16.2 m
Max Speed: 19.3 kn
Cruising Speed: 16 kn
Year of completion: 2010

The Palladium combines sport with a luxurious and communicative spatial concept for physically active passengers. At the unusual size of almost 100 meter, the owner requires dimensions which attracts people of a similar stature. The open main deck is like a club with a swimming pool, numerous seating arrangements and a generous sundeck. The sport garages remind one of the pit lane at Formula 1 races, where mechanics and equipment perform with speed and precision: a complete array of water sport equipment, from jet skis to surfboards for windsurfing, is on board. The exterior of the yacht exudes a timeless elegance but is still unique. The shell is constructed to let natural light into the lower decks. The flowing lines of the exterior continue in the interior, with a minimalist palette of color on the walls, harmonized with the earthy tones on the floors and ceilings .

from left to right: dining area, stairs

main salon

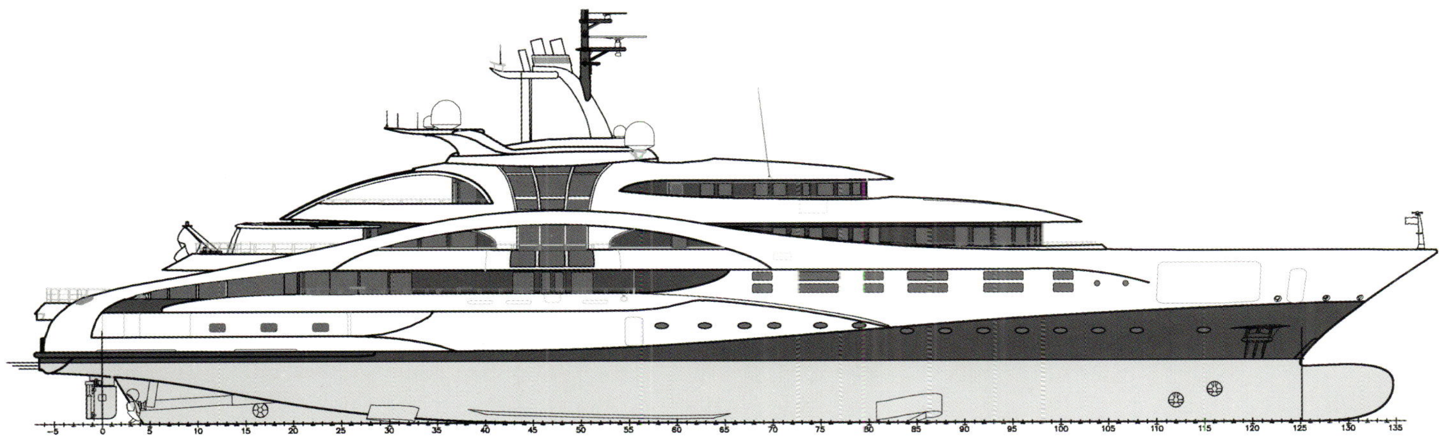

profile

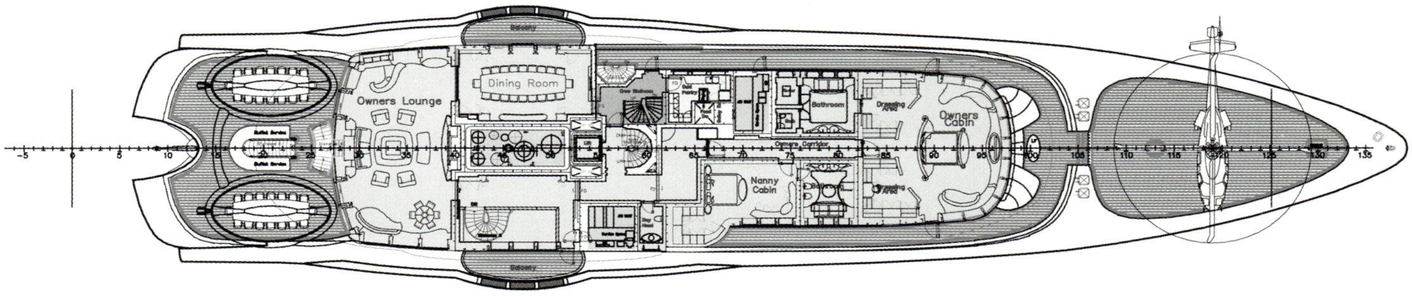

upper deck

cabin

from above to below, from left to right: spa area, bathroom, bathroom detail

from above to below, from left to right: salon, lounge and dining area, stairs

exterior view

Panther 2

Savona, Italy

Interior Design: Luca Dini Design
Exterior Design: Giorgio Vafiadis & Associates
Shipyard: Mondo Marine
Length: 41.56 m
Beam: 8.4 m
Max Speed: 20.5 kn
Cruising Speed: 18.5 kn
Year of completion: 2008

In spite of its 41.6 meter length, the aggressive aerodynamics of the Panther 2 cuts the figure of a sport yacht – and delivers on its promise. Manufactured completely from light alloys, it attains a top speed of 20.5 knots. The design of the interior is no less imposing. The interplay of high quality materials, perfected craftsmanship in the finishing and Italian designer pieces congenially continues the sportive appearance of the exterior into the boat interior. The mega yacht offers luxury and modern diversion for up to twelve guests, who can relax in the salon, tan themselves on the sun deck in an exclusive lounge atmosphere or enjoy a drink at the illuminated bar. From the fleet tender to the jet skis, to the fishing gear, the water toys round out the sportive luxury of the yacht.

from above to below, from left to right: cabin, shower, bathroom detail

Panther 2 at sea

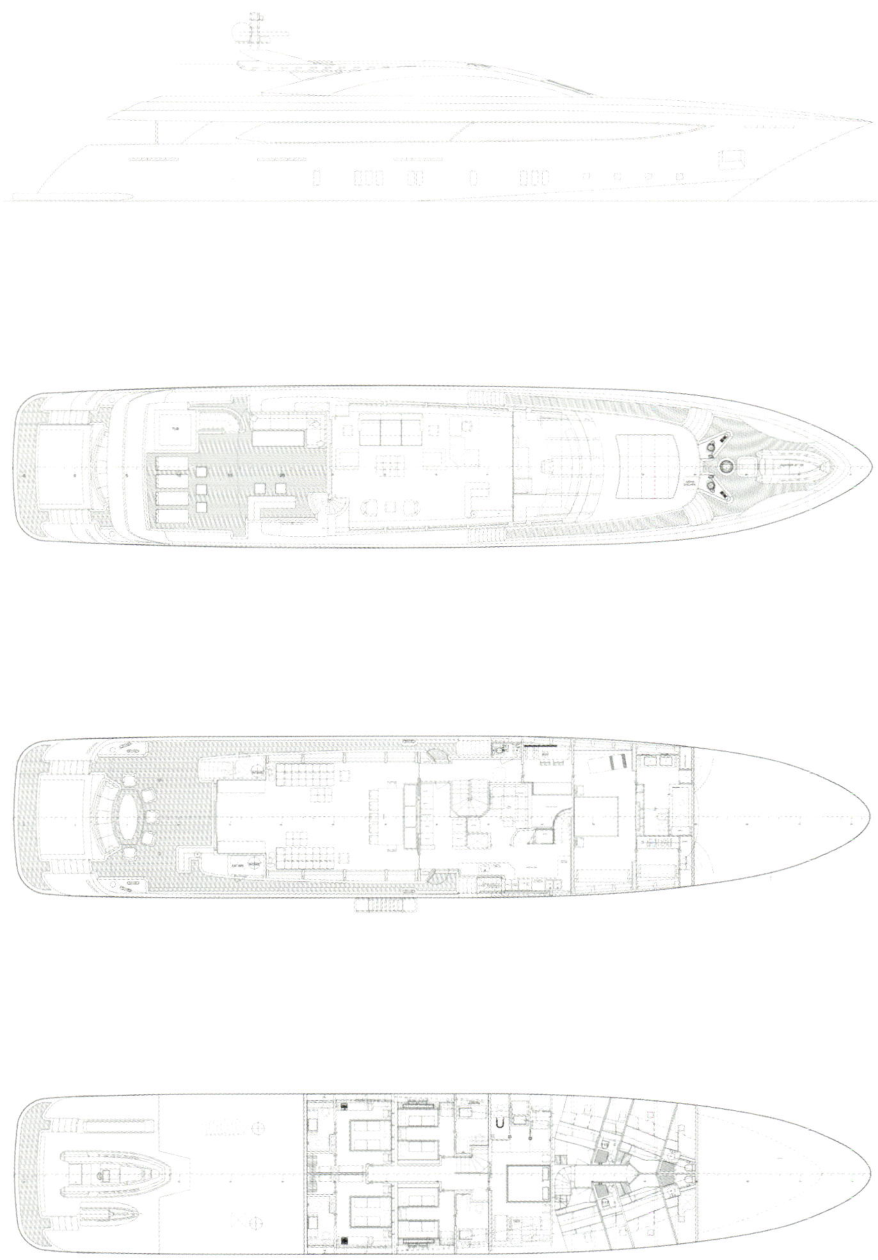

from above to below: profile, bridge deck, main deck, lower deck

from above to below, from left to right: main deck, main salon, dining area

Perle Noire at sea

Perle Noire

Oss, The Netherlands

Design: Frank Laupman, Omega Architects, Gabriel Bernardi
Shipyard: Heesen Yachts
Length: 37.3 m
Beam: 7.5 m
Max Speed: 31 kn
Year of completion: 2010

The hull and superstructure of the 37 meter long Perle Noire are built entirely from aluminium, allowing a top speed of 31 knots. Frank Laupman from Omega Architects designed the slender and elegant exterior lines, which are underscored by the black skin. Gabriel Bernardi from London Interior took charge of the interior design. His inspiration was luxury and the elegance of traditional British craftsmanship. A total of ten passengers have three decks at their disposal, among them the large sun deck with seating area, a bar and spa-pool. In addition to the extra class comfort, great value was placed on functionality. For instance, a convertible corner sofa on the mezzanine deck can be transformed into a bed, ideal for long passages. The special mixture of craftsmanship and high-tech make the Perle Noire into a true pearl.

from above to below, from left to right: dining area, owner's suite, owner's workspace

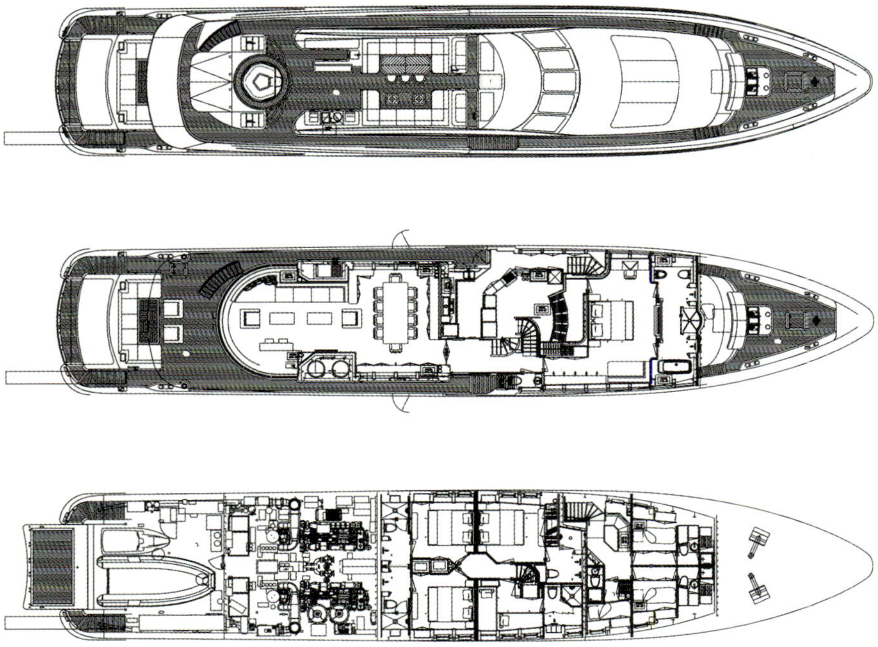

from above to below: sun deck, main deck, lower deck

VIP-cabin

exterior view

stern view

Ra

Design: Adam Voorhees Design
Length: 66 m
Beam: 14.5 m
Max Speed: 22 kn
Cruising Speed: 16 kn

Named after the Egyptian sun god Ra, this yacht uses the power of light in order to establish an unmediated contact with the maritime environment. In spite of its impressive size, the yacht is limited to just three decks, allowing for unheard of dimensions in the interior, with innovative rooms playing on the transition from public to private spaces. The selection of material and clear lines also contribute to blurring the borders between inside and outside. The main deck is a giant multifunctional area. Immediately above it are the guest cabins with a stupendous view. Transparency is the leitmotiv of this vision of a hyper-modern mega yacht, which also measures up to the highest technical demands. The unique form of the Ra guarantees a stable position in the water with minimal fuel consumption. A diesel electric hybrid and azimuth drive complement each other in the machine room.

from above to below, from left to right: lounge, salon, owner's cabin

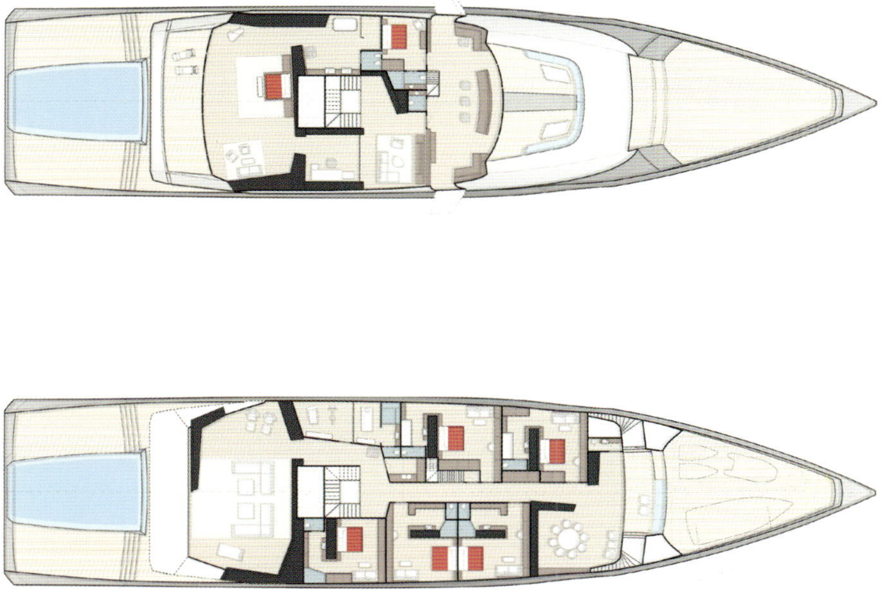

from above to below: owner's deck, guest deck

foyer

from above to below, from left to right: sun deck, upper deck, main deck

exterior view

Sai Ram FB 229

Munich, Germany

Design: Lazzarini Pickering Architetti
Naval architect: Stefano Natucci
Shipyard: Benetti
Length: 52 m
Beam: 10.4 m
Max Speed: 16 kn
Cruising Speed: 15 kn
Year of completion: 2004

In close cooperation with the customers, the architects Lazzarini Pickering created an incomparable perception of space on a motor yacht. Long visual axes and flowing transitions expand the perspectives, while mirrors and glass surfaces appear to raise the ceilings. The palette of materials used is deliberately natural: leather, light woods and natural stone determine the main axes, walls and ceiling panels. Floors and furniture in dark wenge wood define the rooms. The earthy palette of colors of the upholstered furniture creates a warm atmosphere with an oriental touch. Large flat screen monitors connected to a central server in the entertainment area and controlled by touch screen leave no wish unfulfilled.

main salon

from above to below, from left to right: main salon, stairs, floor

from above to below, from left to right: dining area, workspace, owner's bathroom

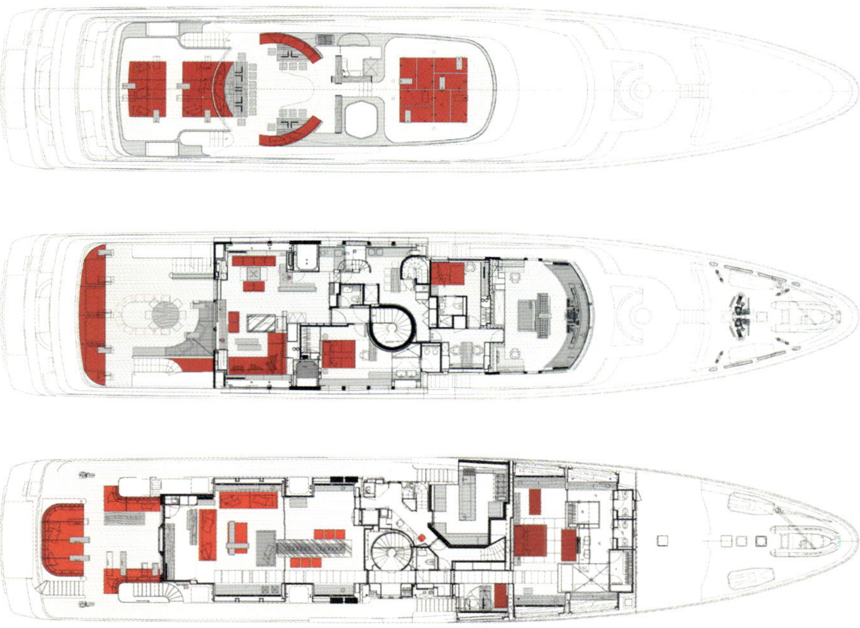

from above to below: sun deck, main deck, lower deck

owner's cabin

from above to below, from left to right: main salon, salon, bar

Sky at sea

Sky

Rijnstraat, The Netherlands

Exterior Design: Omega Architects
Interior Design: Mojo Stumer Associates
Shipyard: Heesen Yachts
Length: 50 m
Beam: 9.6 m
Max Speed: 23 kn
Baujahr: 2010

This yacht is the result of a collaboration between the shipbuilder Heesen and the architecture firm Mojo Stumer. The contrasting disciplines were the occasion for a mutually beneficial cross-pollination cultivation. What emerged was what lured Poseidon himself to the surface of the sea: the Sky, a uniquely graceful yacht, with the highest grade of furnishings. The carbon fiber hull could have been taken from a space vehicle, with navigation and propulsion systems controlled by touch screen. The clear aesthetics inside rely on choice materials like Sevilla leather, Lima limestone, rust-free steel and Starphire glass. No compromises with resilience or durability were made. Hand-picked Macassar ebony covers the walls and on the floors you find in-house fabricated silk carpets. Poseidon is more than a little jealous.

cabin

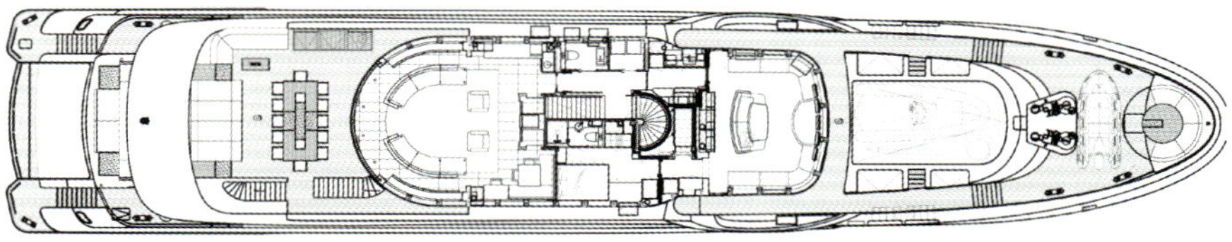

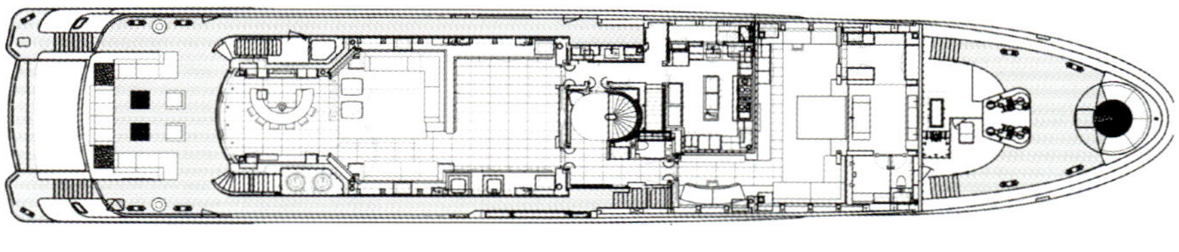

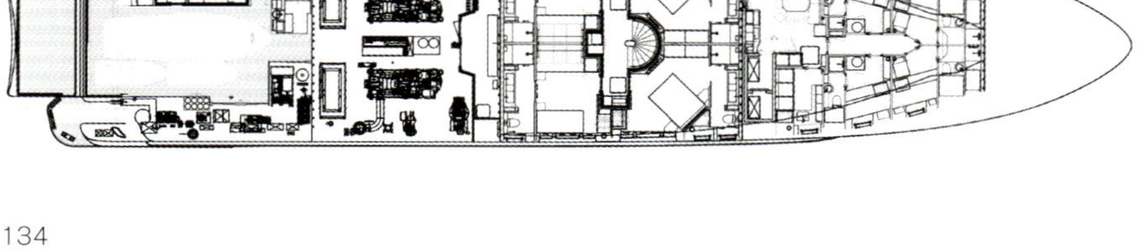

from above to below: upper deck, main deck, lower deck

from left to right, from above to below: workspace, stairs, dining area

sun deck

exterior view

Soja

Fano, Italy

Interior Design: A-Lab Marine
Exterior Design: Stirling Design
Shipyard: Seanest
Length: 25.6 m
Beam: 6.10 m
Max Speed: 30 kn
Year of completion: 2007

The Soja is a serially manufactured 25 meter yacht with a fiberglass hull and superstructure. In collaboration with the new owner, the architects of A-Lab went about fulfilling all his aesthetic desires. For furnishings, they drew on traditional nautical materials: the laying patterns of the teak flooring creates clearly differentiated zones. Mahogany was used for inlays, steel and leather for the edges. The salon and dining area are located on the main deck, with the kitchen separated by a glass sliding wall. On both sides, large panorama windows direct the view to the maritime surroundings. The upper deck is home to the conning tower and living room. A whirlpool, deck chairs and a bar are located on the adjacent sun deck. The lower deck was optimized in order to accommodate the larger cabins, bathrooms and stowage.

from above to below: dining area, main salon

from left to right, from above to below: bathroom, cabin, workspace

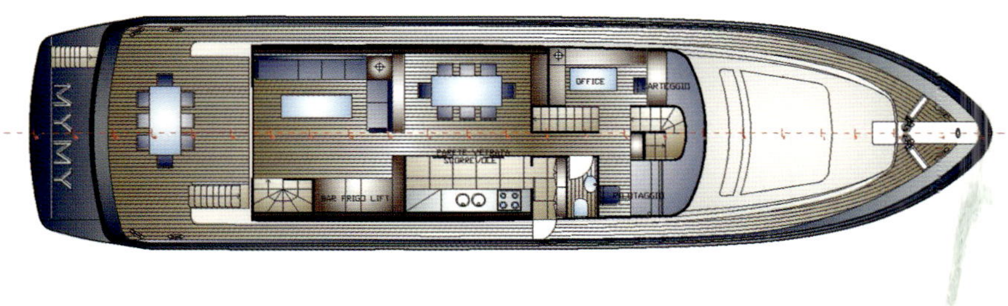

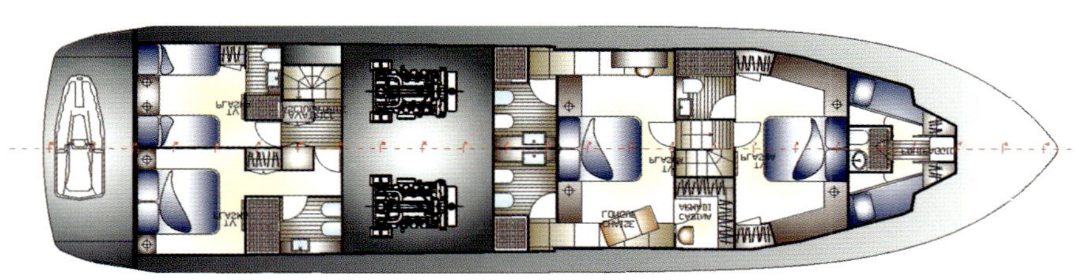

from above to below: flybridge, lower deck

stern view

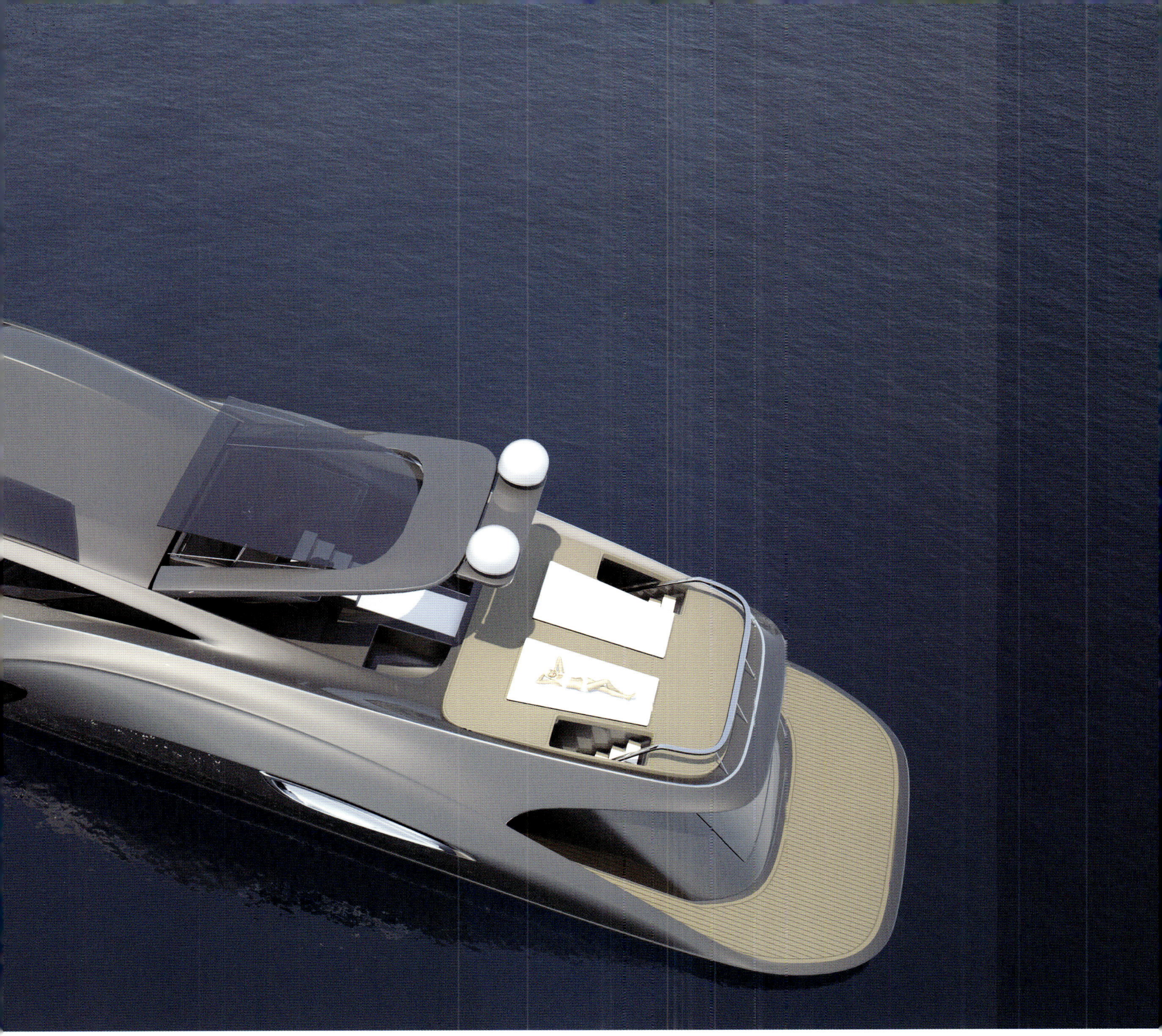

exterior view

Strand Craft

Design: Gray Design
Length: 50 m
Beam: 9 m
Max Speed: 50 kn
Cruising Speed: 9 kn
Completion: 2013

The Strand Craft 166 is a Super Yacht in the uncompromising design of Eduard Gray. Its enormous length and low water line guarantee it an eye opening appearance in every harbor of the world. A vessel preferred by the men among the yacht lovers, because the basic automotive theme lends the yacht a striking male appearance, appealing especially to the lovers of technical sophistication. The conning tower and radar mast can be extended independently of each other. A whirlpool with integrated bar is hidden on the carbon fiber foredeck. Chrome, red leather and LED points of light mark the interior. The two MTU engines which accelerate the vessel to a robust 40 knots are also part of the vessel's automotive heritage. Upon arrival in the harbor the yacht can then be appropriately disembarked in a super car parked in the stern.

from above to below: exterior view, salon

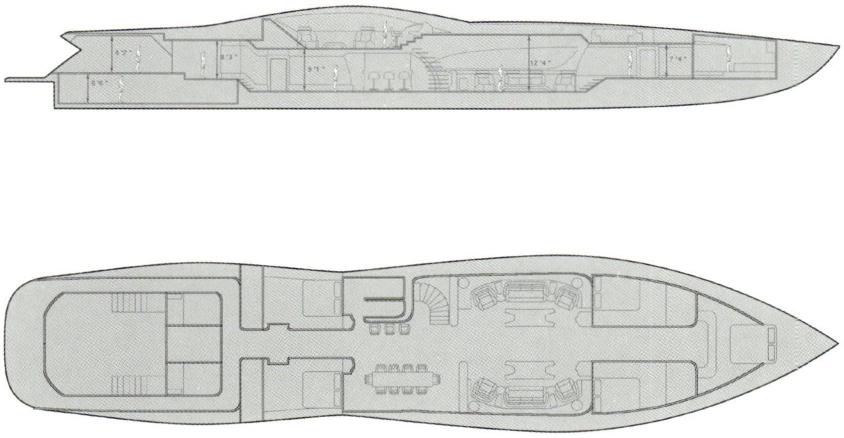

from above to below: section, deck

bow view

exterior view

Streamline 50 m

Savona, Italy

Interior Design: Luca Dini Design and Frédéric Méchiche
Exterior Design: Cor D. Rover
Shipyard: Mondo Marine
Length: 49.3 m
Beam: 9.3 m
Max Speed: 17 kn
Cruising Speed: 16 kn
Year of completion: 2009

In search of timeless beauty, the architect Frédéric Méchiche and the designer Luca Dini selected an elegant black and white contrast for the interior of the Streamline. Modern leather sofas rub shoulders with design classics, while the black lacquered flooring accents the mostly white furniture. The passengers dine with a panorama view beneath a star filled sky on the upper deck, and when a stiff breeze kicks up, the expansive lounge offers refuge with home cinema and surround system. "Streamline" is not just a name for the yacht: Cor D. Rover executed the concept studies and drafted dynamic lines for the steel hull and aluminum superstructures.

from above to below, from left to right: salon, outside living and dining area, look below deck

main salon

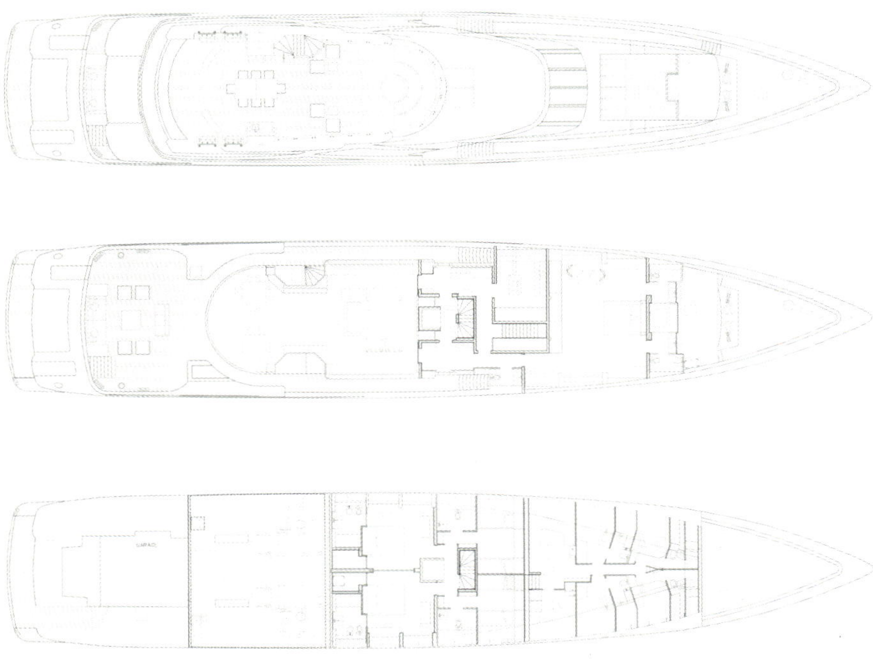

from above to below: sun deck, main deck, lower deck

from above to below, from left to right: outdoor area owner's deck, bow detail, stairs detail

Sycara V at sea

Sycara V

Rendsburg, Germany

Interior Design: Pure Detail
Naval Architect and Exterior Design: Nobiskrug GmbH
Shipyard: Nobiskrug GmbH
Entertainment Systems: Imtech Marine Germany
Length: 68.1 m
Beam: 12.5 m
Max Speed: 17 kn
Cruising Speed: 14 kn
Year of completion: 2010

Sycara V, winner of the "2011 World Super Yacht Award", is a masterpiece of German engineering and craftsmanship. The Nobiskrug Shipyard yacht received the prestigious prize in the 60 to 84 meter class. A low center of gravity, exceptional seaworthiness and the minimal noise level of 35 dB make cruising at a top speed of 17 knots a gentle gliding. While underway the passengers may enjoy unparalleled luxury on five decks and an incomparable entertainment program, from a fitness room, to video and music on demand, to whirlpool and an on-board beach club. In addition to a 25 meter long salon and the library, the interior designed by Craig Beale features many other rooms which qualify the Sycara as a floating palace.

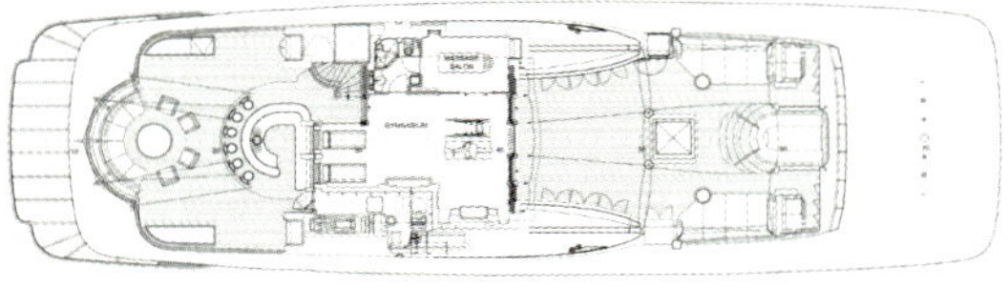

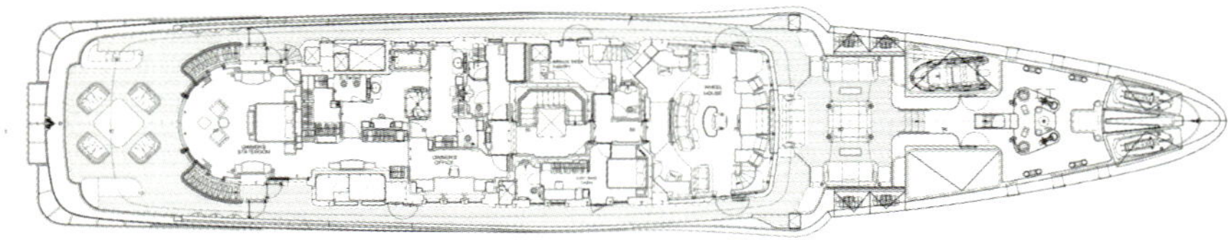

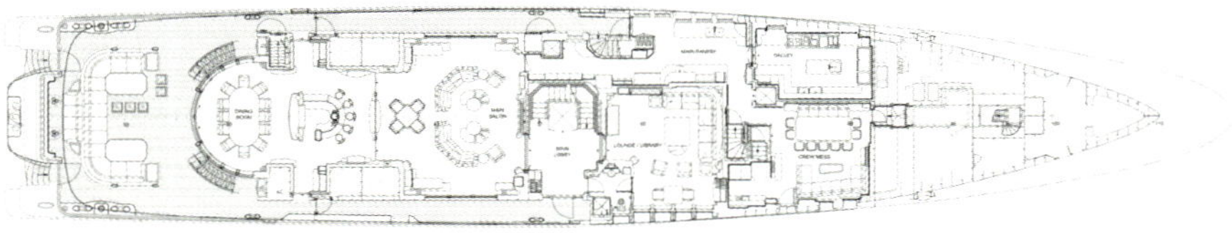

from above to below: sun deck, bridge deck, main deck

main salon

from above to below, from left to right: owner's cabin, owner's workspace, owner's bathroom

from left to right, from above to below: guest cabin, guest bathroom, outdoor area

stern view

from above to below, from left to right: sun deck, jacuzzi, lounge

taTii at sea

taTii

Palma de Mallorca, Spain

Design: Tamsen Yachts
Shipyard: Tamsen Yachts
Length: 40.5 m
Beam: 8.6 m
Max Speed: 20 kn
Year of completion: 2009

The Indian name taTii means "beach" and in fact, this 41 meter long yacht has the charm of a lonely island. The voluminous hull houses three decks, while managing to maintain a sportive and elegant appearance. Fine materials like Zebrano wood and ebony rule the interior, with white leather providing an agreeable contrast. Clear lines run through the room layout and the furniture style, from the loveseat to the deck chairs. Another highlight is the sound system, which thanks to the extraordinary acoustic insulation can also be enjoyed at top speed. When the sun deck is transformed into a dance floor or the stillness of the night is contemplated through the broad sun roof, then the taTii leaves behind every island in its wake.

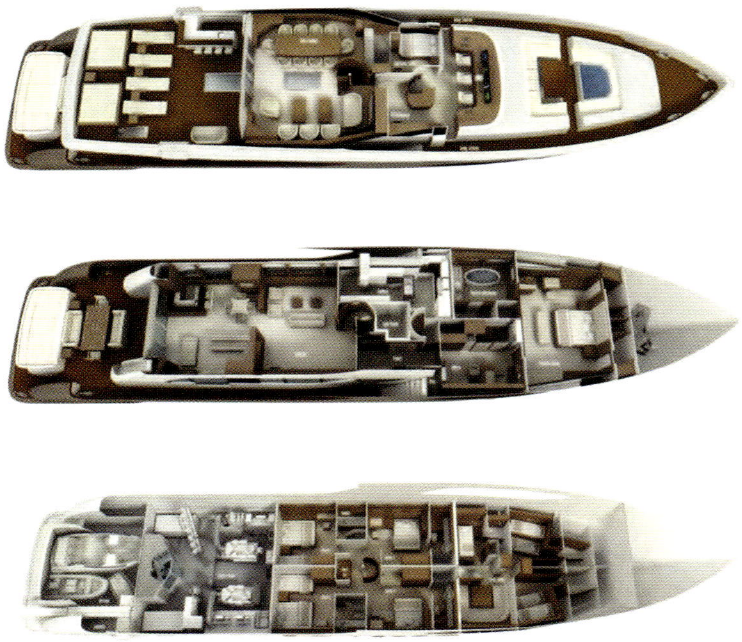

from above to below: sun deck, main deck, lower deck

from left to right, from above to below: guest cabin, hall, guest bathroom

from left to right, from above to below: workspace, owner's bathroom, owner's cabin

from above to below, from left to right: main salon, dining area, exterior view

Tribù at sea

Tribù

Savona, Italy

Interior Design: Studio Lissoni Associati
Exterior Design: Luca Dini Design & Massimo Vignelli
Shipyard: Mondo Marine
Length: 50.5 m
Beam: 10.4 m
Max Speed: 15 kn
Cruising Speed: 13 kn
Year of completion: 2007

When a prominent Italian fashion designer decides in favor of a yacht, the standards of design and workmanship are correspondingly high. The Tribù was the wish of no less than Luciano Benetton. The 50 meter long expedition ship won many prizes, among them for its overall concept as a marine vessel and for its unconventional lines. Aluminum superstructures are mounted on a steel hull, which lend the yacht its characteristic silhouette. Benetton not only evidences good taste, he also demonstrates his environmental consciousness. Tribù is the first yacht to receive the prestigious Green Star Certificate of the Italian Naval Registry. The award only goes to yachts which minimize their ecological footprints with innovative engineering.

from left to right: hall, salon

cabin

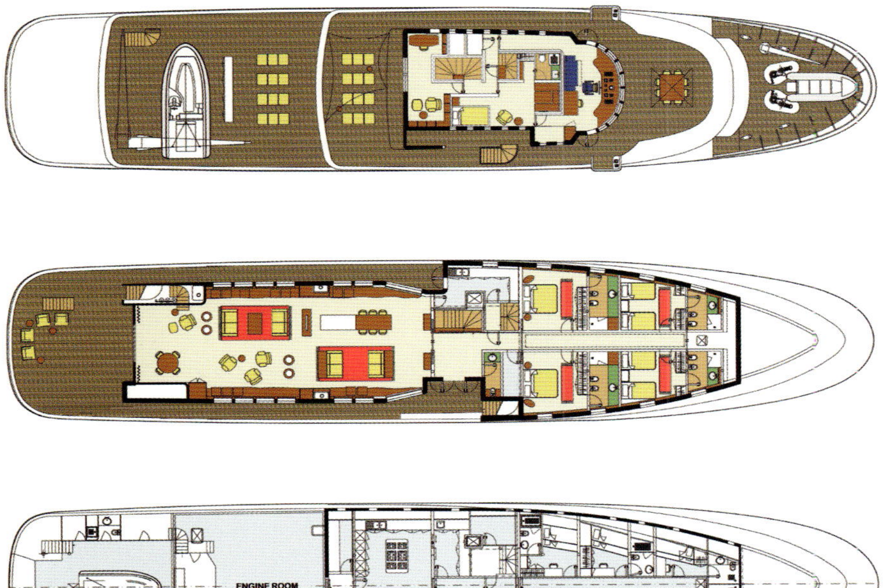

from above to below: upper deck, main deck, lower deck

exterior view

118 Wallypower at speed

118 Wallypower

Monte Carlo, Monaco

Design: Wally, Lazzarini Pickering Architetti, Luca Bassani Design
Shipyard: Intermarine
Length: 36 m
Beam: 9 m
Max Speed: 65 kn
Cruising Speed: 45 kn
Year of completion: 2002

When technology meets design with the Wally pure path approach, the result is a unique, very fast, strong and sexy super yacht. The technical solutions blend with the styling solutions producing unique and marked characteristics that contribute to the yacht breakthrough appeal and performance, like the vertical bow, the air inlets, and the deck superstructure. The building technology provides for a yacht with the strength of a mine-sweeper combined with the lightness of a jet-fighter. The 16,800 HP gas turbines generate 65 knots of maximum speed. The deck, cockpit, navigation and saloon areas are one continuous element to enjoy open air living. The 118 Wallypower marked a turning point in naval architecture and in contemporary design history.

from above to below, from left to right: bow view, folding passerelle / swimming ladder, opening bulwarks

from above to below: profile, aft deck

from above to below, from left to right: living area under the superstructure, dining area with a retractable table

from left to right: owner's stateroom, owner's bathroom

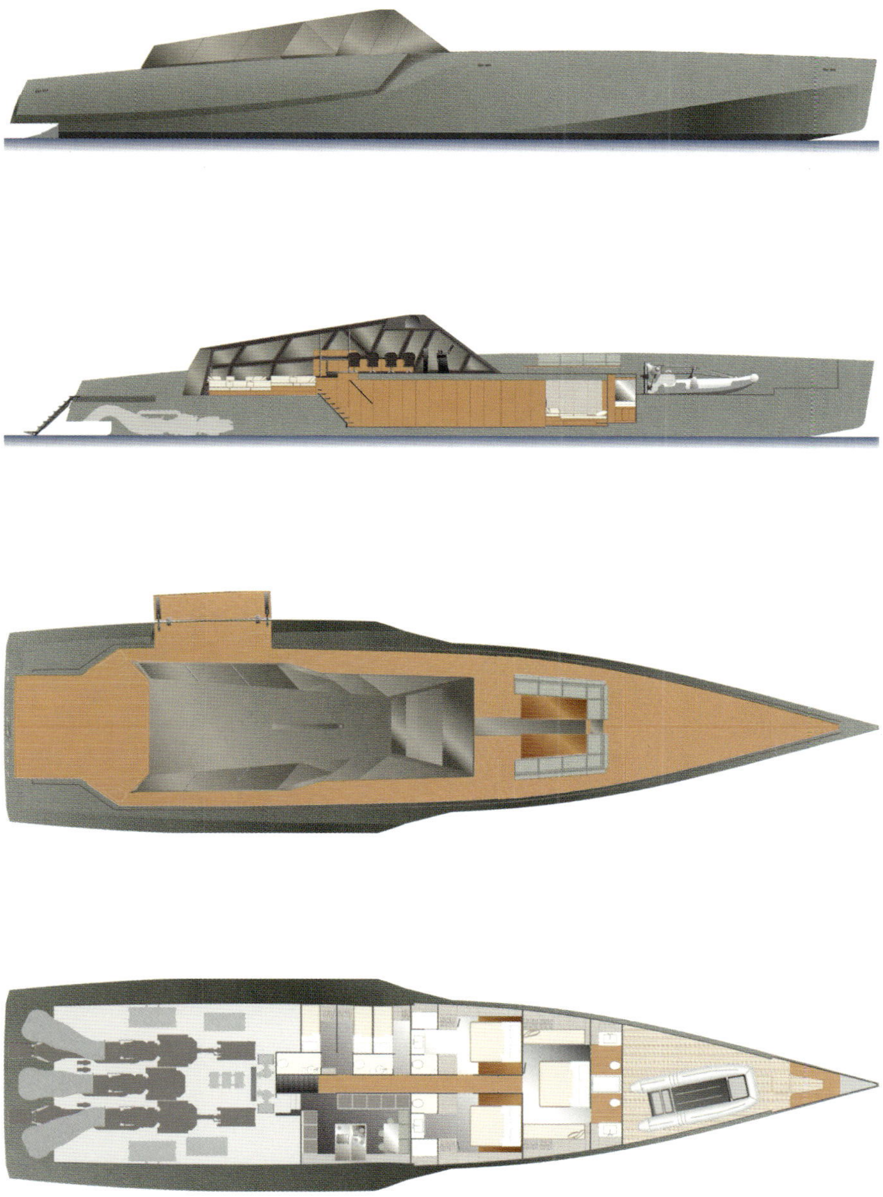

from above to below: profile, section, service and night areas, accommodation

from above to below, from left to right: opened roof, closed roof, afterdeck

exterior view

WHY 58x38

Design and development:
WHY design Team (Luca Bassani, Pierre-Alexis Dumas, Gabriele Pezzini, Mauro Sculli with Claudio Cicconetti)
Interior Design:
Denis Montel, RDAI
Naval Architect:
Roar Ramde - Ramform
Length: 58 m
Beam: 38 m
Max Speed: 14 kn
Cruising Speed: 12 kn

WHY is a new type of motor yacht redefining the art of living on the sea, based on everybody's dream to own an Island: a place to live in complete freedom, with no constraints, with the independence that only the self sufficiency can provide. The concept of the moving Island is developed with the most advanced technologies of sustainable energies. WHY was conceived by Hermès and Wally, assembling a team of dedicated talents in design, engineering, and naval architecture. The tested Norwegian Ramform hull characterised by unprecedented stability and volume was adapted to the project. Other important innovations introduced by WHY: patios generously flooding the interior with natural light, photovoltaic panels on the glass hull sides, as well as on the glass roof opening like Venetian blinds, a forward end swimming pool, and a huge aft deck beach. The groundbreaking design comes in two sizes: WHY 58x38 (58m long and 38m wide) with four decks offering almost 4,000 m^2 of guest area for 16 guests, and WHY 37x24 (37m long and 24m wide) with three decks providing 1 000 m^2 living area for 10 guests.

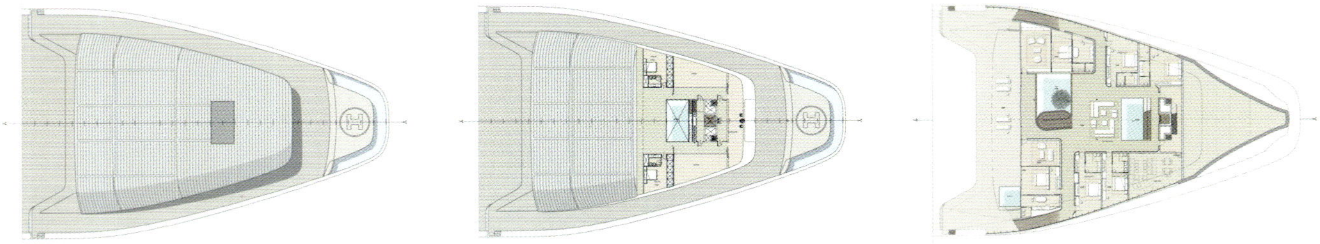

from left to right: top view, wheelhouse deck, owner deck

from left to right, from above to below: dining area, guest area with patio, living area

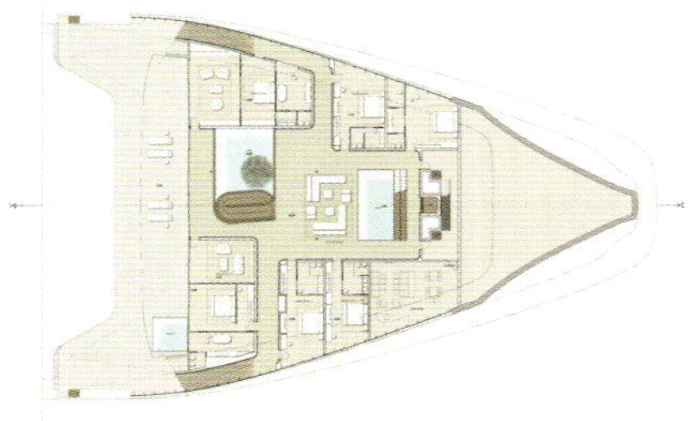

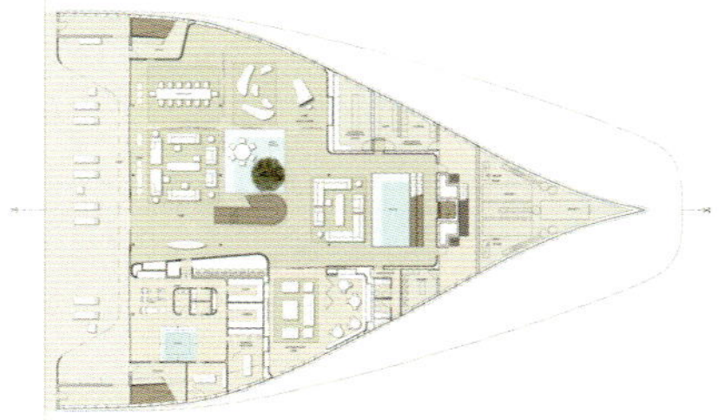

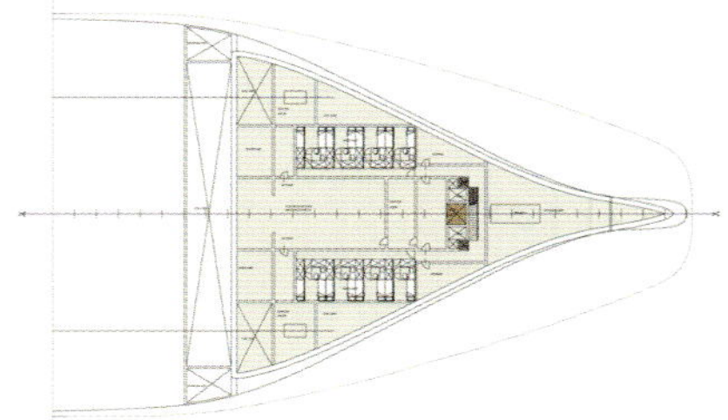

from left to right: upper deck, main deck, lower deck

from above to below: owner's deck, owner's large suite

from above to below, from left to right: cabin, kitchen, bathroom

exterior view

Wider 42'

Castelvecchio, Italy

Design: Fulvio De Simoni
Shipyard: Wider S.R.L.
Length: 12.9 m
Beam: 3.5 m
Beam open configuration: 6.6 m
Max Speed: 53 kn
Cruising Speed: 40 kn
Year of completion: 2011

From adrenaline to rest – the new Wider 42' offers a successful blend of both. The powerful day cruiser is up to any challenge. Taking off from an idea by Tilli Antonelli, the designer Fulvio De Simoni designed this astonishingly many-sided yacht. Flexible extensions on both sides of the hull deploy in just twelve seconds. They double the usable deck surface, while creating additional stability.

Modern demands for fuel economy are also more that fulfilled. The exceptional weight to performance ratio allows the yacht to attain a top speed of 53 knots with minimal fuel consumption. With a cruising speed of 40 knots the yacht has a range of 400 sea miles, an unheard of value for a vessel of this class.

view in the open mode

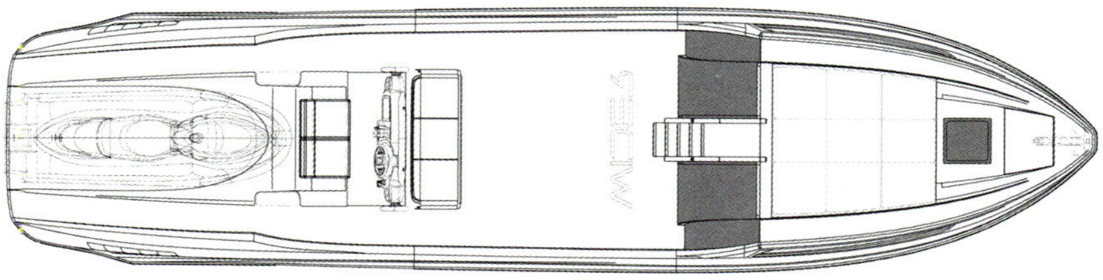

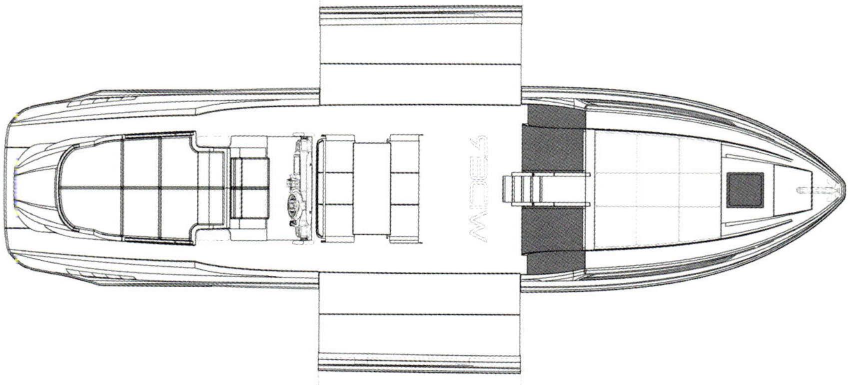

from above to below: closed mode, open mode

exterior view

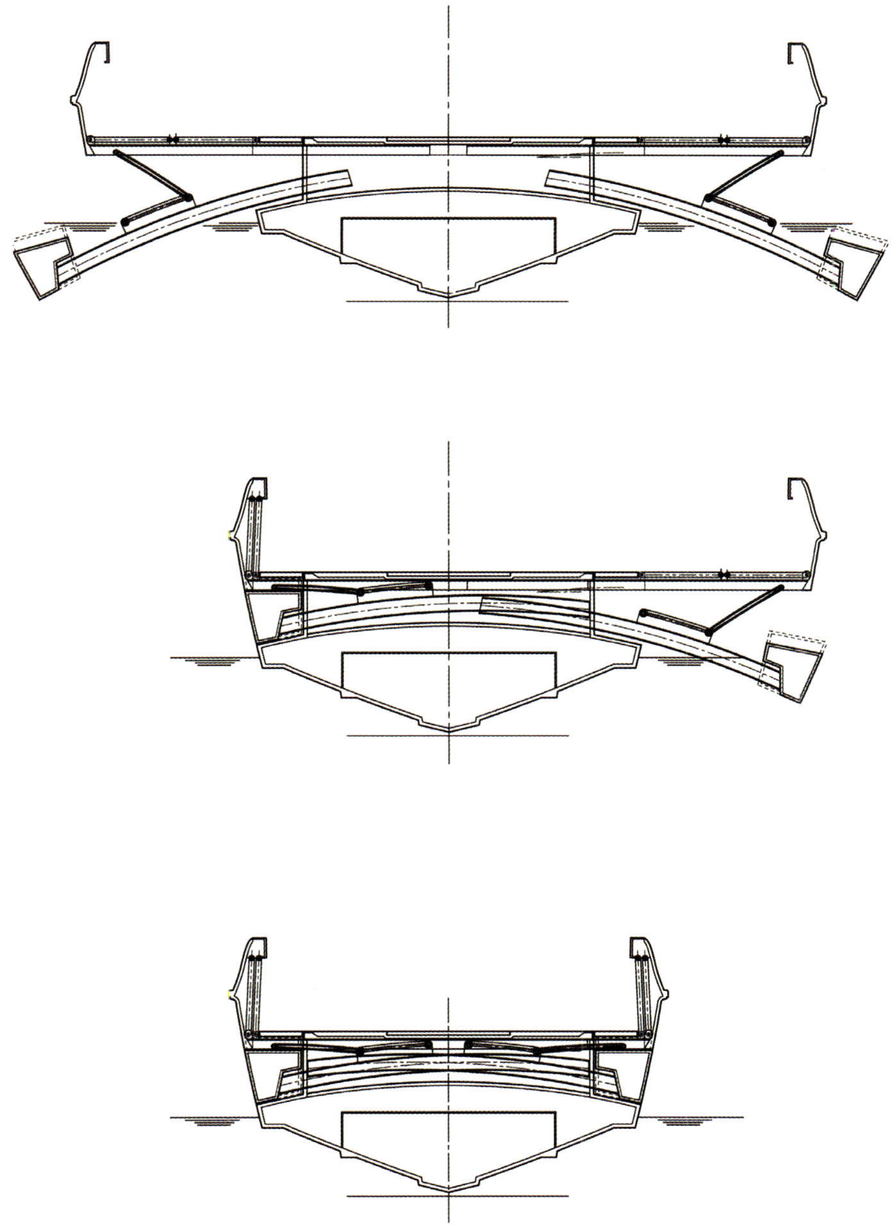

from above to below: section open mode, section half-open mode, section closed mode

from left to right: cockpit, exterior view

XSR 48 at sea

XSR 48

Lymington, United Kingdom

Design: Redman Whiteley Dixon, Fabio Buzzi
Shipyard: XS Marine Group
Length: 14.6 m
Beam: 3.1 m
Max Speed: 85 kn
Cruising Speed: 50 kn
Year of completion: 2008

The XSR 48 from XSMG transmits the feeling of Formula 1 on the high seas. The exclusive powerboat is being produced in a limited series of 100. Renowned yacht designers like Fabio Buzzi and Redman Whiteley Dixon were involved in the development. They were inspired by a sports car, the Bugatti Veyron. The result is a powerboat which combines the power and strength of a racing car with the design of a super yacht. Formula 1 technology has been employed, from the carbon fiber hull to the steering station. Two twin turbine diesel engines with a formidable 1,600 horsepower accelerate the XSR 48 to an impressive 85 knots, in spite of the weight of 7.5 tons. That means the 14.6 meter long yacht is no longer a lightweight, but it glides stably enough over the water to show that styling and performance make an ideal combination.

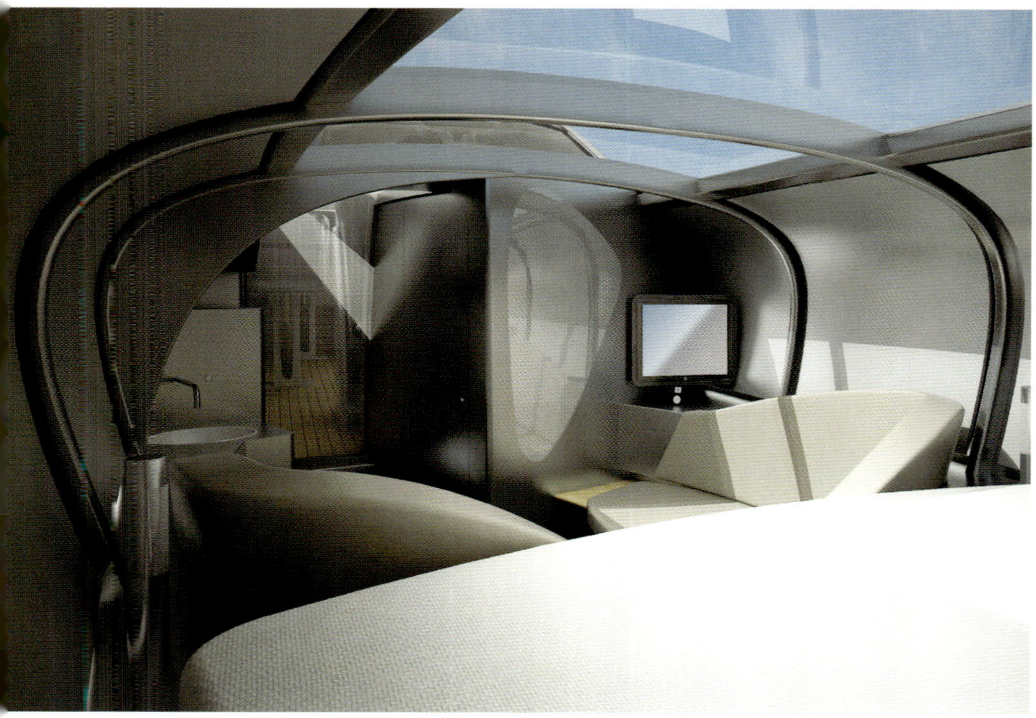

from above to below, from left to right: cabin, bathroom

exterior view

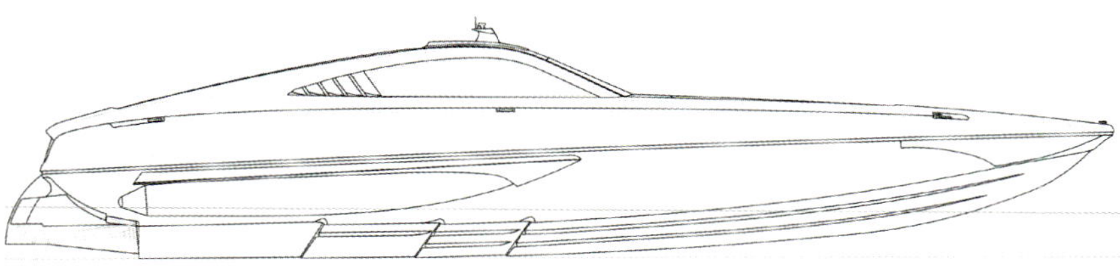

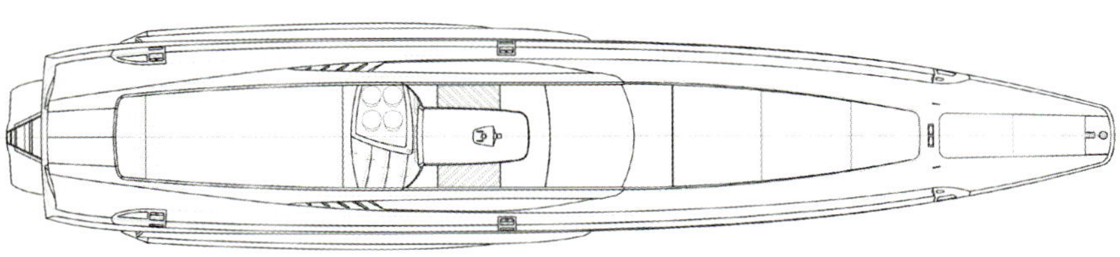

from above to below: profile, deck

from above to below: open-air area, exterior view

from above to below, from left to right: 4H at sea, open-air area, gull-door

4H
Viareggio, Italy

Design: Francesco Paszkowski Design
Shipyard: Sanlorenzo Spa
Length: 40.8 m
Beam: 7.9 m
Max Speed: 28 kn
Year of completion: 2007

Weightlessness is the basic theme of the 4H. This becomes clear when the dynamic lines of the yacht makes contact with the water, in the right mix of aggressiveness and suppleness. The free floating architecture also pervades the interior rooms, which feature the comfort one expects from a yacht of this size. Thanks to the sophisticated lighting, all the furnishings seem to float, although heavy materials like leather and black lacquer prevail. Movable curtains make possible a flexible division of the cabins and create private spaces of refuge. Its many facets and a love of detail make this yacht and emotional wellness boat, whereby the term here does not refer to classic features like a sauna or gym, but rather the feeling for living that life on board this vessel brings with it.

from above to below, from left to right: main salon, stairs, lower deck terrace

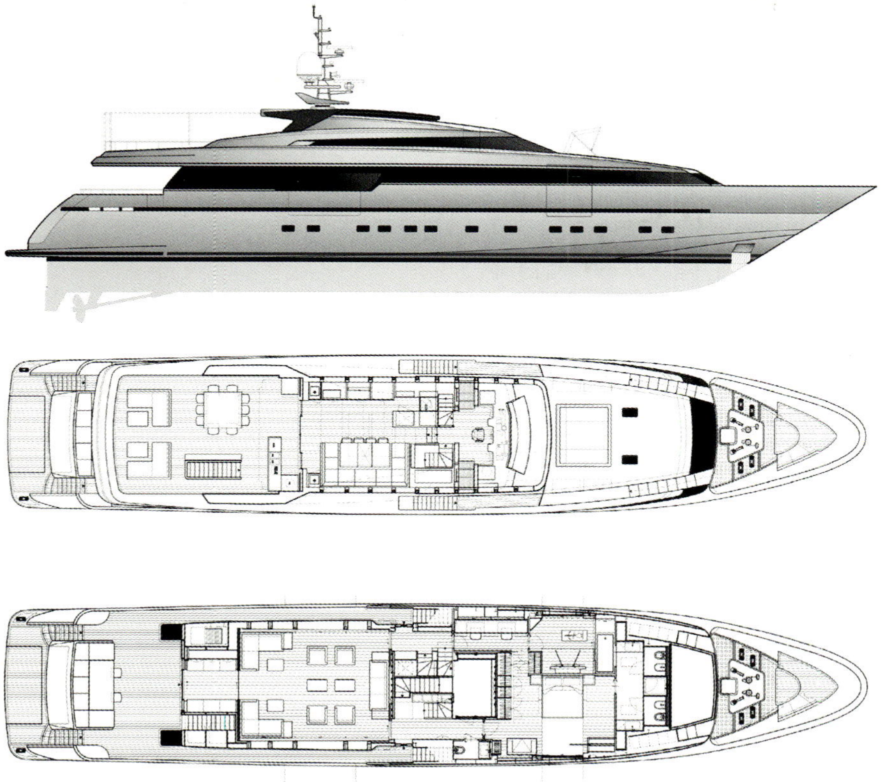

from above to below: profile, flybridge, main deck

owner's cabin

92' Duchessa at sea

from above to below, from left to right: side view, oblique view, exterior area

92' Duchessa

La Spezia, Italy

Design: Officina Italiana Design, Ferretti Group AYT and Centro Stile Ferretti Group
Shipyard: Riva (Ferretti Group)
Length: 28.2 m
Beam: 6.8 m
Max Speed: 28 kn
Cruising Speed: 25 kn

Since 1842, the name Riva has stood for elegance, luxury, craftsmanship and a love of detail in shipbuilding. With the 92' Duchessa, the shipyard has shown once again how one can combine the most modern technology and classic design into a floating masterpiece. Light marble sets the tone below deck, while the warm walnut shades of the hand-sewn leather insets of the furniture and ceiling blend with the white varnished wood flooring. The interior thereby exudes an atmosphere of timelessness and, at the same time, comfort. Racy curves distinguish the yacht from the outside. The conning tower fits seamlessly into this profile. It can also be equipped with a fiberglass hardtop for added protection from the sun. These are only a few of Riva's many well thought-out traditional shipbuilding details which are cherished by the lovers of yachting.

from above to below, from left to right: salon with dining area, cabin, bathroom

from left to right, from above to below: details

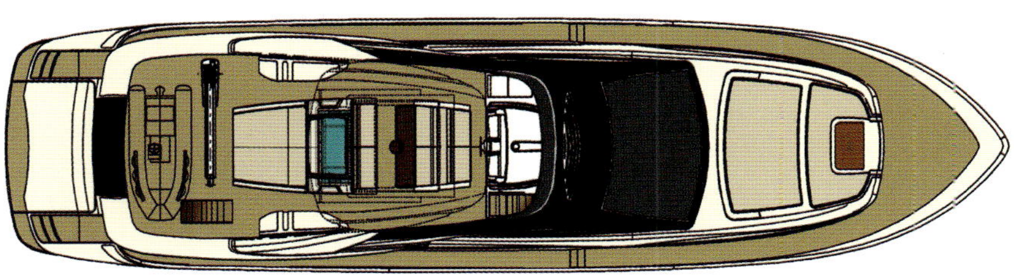

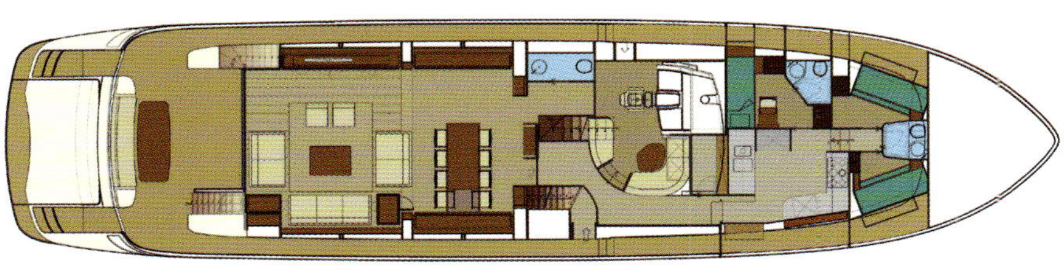

from above to below: flybridge, main deck

Designers' Index

AB Yachts
>> 8, 14

Via Dorsale13
54100 Massa (Italy)
T +39.0585253497
F +39.0585251650
info@abyachts.com
www.abyachts.it

Adam Voorhees Design
>> 122

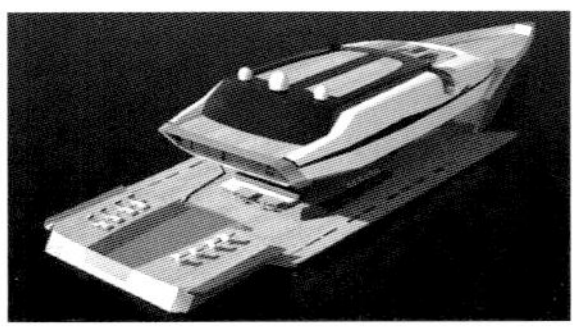

California (USA)
T +1.8187313157
hello@adamvoorhees.com
www.adamvoorhees.com

A-Lab Marine
>>56, 90,136

Via Stacchetti 5
16148 Genova (Italy)
T +39.0103071501
F +39.0108631912
info@alabmarine.it
www.alabmarine.it

Arcadia Yachts
>> 26

Via Terragneta 90
80058 Torre Annunziata
Naples (Italy)
T +39.08119554898
F +39.08119554894
info@arcadiayachts.it
www.arcadiayachts.it

Art-Line Interiors
>> 22

Hulkensteinseweg 35
6812 DB Arnhem
(The Netherlands)
T +31.264436703
F +31.264459182
artline1@chello.nl
www.artline-design.com

Art of Kinetik
>> 52

Kopitareva Gradina 3
11000 Belgrade (Serbia)
T +381.113752310
F +381.113752314
info@artofkinetik.com
www.artofkinetik.com

Aquos Yachts
>> 36

268 Camelot Drive
Tavernier, FL 33070 (USA)
T +1.9547709624
www.aquosyachts.com

Baglietto Design Department
>> 22

Piazza Stefano Baglietto 3
17019 Varazze (Italy)
T +39.01995901
F +39.01996515
info@baglietto.com
www.baglietto.com

Bilgin Yachts
>> 78

Ic Kumsal Mevkii Bizim Sokak 45
Kucukcekmece
Istanbul (Turkey)
T +90.2125996353
F +90.2125996243
info@bilginyacht.com
www.bilginyacht.com

Boris Levyant and Christiaan Bolinger for Acico Yachts
>> 18

Voorland 16
1601 EZ Enkhuizen
(The Netherlands)
T +31.228351900
F +31.228351909
info@acico-yachts.com
www.acico-yachts.com

CNC (Cantieri Navali Crotone)
>> 40

Nuova Zona Industriale
Via G. Mercalli
88900 Crotone (Italy)
T +39.0962930791
F +39.0962938662
administration@feliniyachts.com
www.feliniyachts.com

Cor D. Rover
>> 144

Buitenhaven 29
2965 AD Nieuwpoort
(The Netherlands)
T +31.184609333
F +31.184609444
www.cor-d-rover.com

De Birs Yachts
>> 68

Orlyplein 10 - Floor 24
1043 DP Amsterdam
(The Netherlands)
T +31.208080028
sales-mrktg@debirsyachts.com
www.debirsyachts.com

Diana Yacht Design
>> 94

Ampèrestraat 13-A
1817 DE Alkmaar
(The Netherlands)
T +31.725200155
F +31.725209191
info@dianayachtdesign.com
www.dianayachtdesign.com

Egg and Dart Design Corporation
>> 68, 72

Zeppelinstraße 61
81669 Munich (Germany)
T +49.897673650
F +49.8976736525
studio@egganddart.de
www.egganddart.de

Fabio Buzzi
>> 178

Via Provinciale 73
23841 Annone Brianza (Italy)
T +39.0341260105
F +39.0341260108
mail@fbdesign.it
www.fbdesign.it

Ferretti Group Centro Stile, AYT
>> 186

Via Ansaldo 5/7
47122 Forlì (Italy)
T +39.0543787511
fulvia.venturi@ferrettigroup.com
www.ferrettigroup.com

Fipa Group
>> 8, 14

Via Marina di Levante 12/14/16
55049 Viareggio (Italy)
T +39.058438191
F +39.05843819333
info@fipagroup.com
www.fipagroup.com

Foster+Partners
>> 98

Riverside
22 Hester Road
London SW11 4AN (UK)
T +44.2077380455
F +44.2077381107
info@fosterandpartners.com
www.fosterandpartners.com

Guida Engineering + Design
>> 26

C.so V. Emanuele 140
80121 Naples (Italy)
T +39.0816581677
F +39.0816581694
f.guida@guidadesign.it
www.guidadesign.it

Francesco Paszkowski Design
>> 182

Via S. Maria a Marignolle 69
50124 Florence (Italy)
T +39.055229519
info@paszkowskidesign.it

Frédéric Méchiche for Mondo Marine
>> 144

Lungomare Matteotti 6
17100 Savona (Italy)
T +39.019828516
F +39.019823418
mondomarine@mondomarine.it
www.mondomarine.it

Fulvio De Simoni for Wider s.r.l.
>> 174

Viale Marche 2/4
61040 Castelvecchio di Monte
Porzio (Italy)
T +39.0721956077
F +39.0721956001
info@wider-y.it
www.wider-y.it

Gabriel Bernardi
>> 118

59 Pimlico Road
London SW1W 8NE (UK)
T +44.2072590960
F +44.2072590961
info@cooteandbernardi.com
www.cooteandbernardi.com

Giorgio Vafiadis & Associates
>> 114

Viale della Astronomia 8
00144 Rome (Italy)
T +39.065920066
F +39.065920883
vafiadis@interin.it

Gray Design
>> 140

Ornvagen 52
46833 Vargon (Sweden)
T +46.705185450
info@graydesign.se
www.graydesign.se

Gregory C. Marshall
>> 36

1571 West Burnside Road
Victoria, BC V9E 2E2 (Canada)
T +1.2503889995
F +1.2503884260
info@gregmarshalldesign.com
www.gregmarshalldesign.com

Guido De Groot
>> 14

Hogewoerd 122
2311 HT Leiden (The Netherlands)
T +31.715663040
www.guidodegroot.com

HanseYachts
>> 48

Salinenstraße 22
17489 Greifswald (Germany)
T +49.383457920
F +49.3834579281
www.hanseyachts.com

Heesen Yachts
>> 132

Rijnstraat 2
5347 KL Oss (The Netherlands)
T +31.412665544
F +31.412665566
press@heesenyachts.nl
www.heesenyachts.nl

Hot Lab Yacht & Design
>>40, 78

Via Cevedale 7
20158 Milan (Italy)
T +39.0240093293
F +39.02700509446
info@hotlab.it
www.hotlab.it

Lazzarini Pickering Architetti
>> 82, 126, 162

Via Cola di Rienzo 28
00192 Rome (Italy)
T +39.063210305
F +39.063216755
info@lazzarinipickering.com
www.lazzarinipickering.com

Luca Dini Design
>> 114, 144, 158

Via Venezia 4
50121 Florence (Italy)
T +39.0555059379
F +39.0554089050
info@lucadinidesign.com
www.lucadinidesign.com

Michael Leach Design
>> 108

Church Lane
East Boldre, New Forest
Hampshire SO42 7WS (UK)
T: +44.1590612908
F: +44.1590612735
mail@michaelleachdesign.com
www.michaelleachdesign.com

Mojo Stumer Associates
>> 132

14 Plaza Road
Greenvale, NY 11549 (USA)
T +1.5166253344
F +1.5166253418
web@mojostumer.com
www.mojostumer.com

Mulder Design
>> 72

Benedeneind Zuidzijde 289B
3405 CK Benschop
(The Netherlands)
T +31.348439680
F +31.348439681
info@mulderdesign.nl
www.mulderdesign.nl

Navirex
>> 56

Via Galata 39/6
16121 Genova (Italy)
T +39.0105761129
F +39.0105452154
navirex@tin.it
www.navirex.it

Nobiskrug GmbH
>> 148

Kieler Straße 53
24768 Rendsburg (Germany)
T +49.43312070
info@nobiskrug.com
www.nobiskrug.com

Officina Italiana Design
>> 186

Via San Tomaso 27
24121 Bergamo (Italy)
T +39.035245765
F +39.035240937
officina@italianadesign.it
www.italianadesign.it

Olivier F. van Meer Design
>> 74

Oosterhavenstraat 39
1601 KX Enkhuizen
(The Netherlands)
T +31.653311089
info@vanmeerdesign.nl
www.vanmeerdesign.com

Omega Architects
>> 118,132

Klepperheide 22
6651 KM Druten (The Netherlands)
T +31.487518889
omega@omega-architects.com
www.omega-architects.com

Patrick Banfield and Allseas Design
>> 48

sales@fjordboats.com.au
www.fjordboats.com.au

Perini Navi & Picchiotti
>> 44

Via Coppino N. 114
Viareggio (Italy)
www.perininavi.it

Philippe Briand
>> 44

60 Sydney Street
London SW3 6PS (UK)
T +44.2073512700
F +44.2073498500
yachts@philippebriand.com
www.philippebriand.com

Pure Detail
>> 148

Melville (Australia)
T +61.893308350
puredetail@bigpond.com.au

Puresang and Will Erens
>> 32

2000 Antwerp (Belgium)
T +32.474690402
info@pure-sang.com
www.pure-sang.com

RDAI
>> 168

13 Rue du Mail
75002 Paris (France)
T +33.142600482
F +33.140150429
contact@renadumas.com
www.renadumas.com

Redman Whiteley Dixon Ltd.
>> 62, 178

The Old Electric Light Station
Beaulieu
Brockenhurst
Hampshire SO42 7YF (UK)
T +44.1590611300
F +44.1590611301
studio@rwd.co.uk
www.rwd.co.uk

Schnaase Interior Design
>> 94

Hafentor 2
20459 Hamburg (Germany)
T +49.404107721/27
F +49.40417725
info@schnaase.de
www.schnaase.de

Schöpfer Yachts
>> 104

160 Commonwealth Avenue
Suite 5L
Boston, MA 02116 (USA)
contact@schopferyachts.com
www.schopferyachts.com

Stirling Design
>> 90, 136

6 Rue Eaboneau
44100 Nantes (France)
T +33.240957945
F +33.240957946
contact@stirlingdesign.fr
www.stirlingdesign.fr

Studio Linse
>> 62

Keizersgracht 534
1017 EK Amsterdam
(The Netherlands)
T +31.206754798
F +31.206719068
info@studiolinse.com
www.studiolinse.com

Studio Lissoni Associati
>> 158

Via Goito 9
20121 Milan (Italy)
T +39.026571926/42
F +39.026571918
info@lissoniassociati.it
www.lissoniassociati.com

Sunseeker Design
>> 86

Kajen 6-8
20459 Hamburg (Germany)
T +49.40555050
F +49.4055505150
info@sunseeker.de
www.sunseeker.de

Elbchaussee 223
22605 Hamburg (Germany)
T +49.40555016660
F +49.405550166616
info@tamsen.com
www.tamsen.com

Vignelli Associates
>> 158

130 East 67th Street
New York, NY 10021 (USA)
T +1.2127370538
F +1.2127444846
info@vignelli.com
www.vignelli.com

Vitruvius Yachts
>> 44

60 Sydney Street
London SW3 6PS (UK)
T +44.2073512728
F +44.2073498500
info@vitruviusyachts.com
www.vitruviusyachts.com

Wally
>> 162, 168

6, Avenue Albert II
Monte-Carlo
MC 98000 (Monaco)
T +377.93100093
F +377.93100094
info@wally.com
www.wally.com

Picture Credits

Ardash, Mohamed	68–71
Baldi, Maurizio	182, 183, 184 b. r.
Bianchi, Emilio	118–121
Cianci, Bruno	78–81
Cocchi , Alberto	56–61
Grugnola, Guido	166, 167
Holthuis, Dick	24, 25
Malgarini, Giovanni	40–43
Martin-Raget, Gilles	162–165
McKenna, Tim	36 b., 37 a. l., 37 r.
Miras, Ricardo	52–55
Piazza, Matteo	82–85, 128–131
Rabinowitz, Neil	36 a., 37 b. l., 38, 39
Rizzi, Laura	184 a., 184 b. l.,185
Tangram 3DS	104–107
Thomas, Bruce	94–97
Sahm, Stephan	72–77
Sargentini, Giuliano	44–47
WHY/ Artefactory Lab	168–173
Young, Nigel /Foster+Partners	98–103

Cover: AB Yachts Fipa Group
Backcover from above to below, from left to right:
Sunseeker International Limited, Nigel Young /
Foster+ Partners, Luca Dini Design

All other pictures were made available by the architects, designers or shipyards.